research and exposition in mathematics

Edited by

Karl H. Hofmann
Fachbereich Mathematik
Technische Hochschule
Schloßgartenstr. 7
D - 6100 Darmstadt
Fed. Rep. Germany

Rudolf Wille
Fachbereich Mathematik
Technische Hochschule
Schloßgartenstr. 7
D - 6100 Darmstadt
Fed. Rep. Germany

Titles in this Series

1 R.T.Rockafellar: The theory of subgradients and its applications to problems of optimization. Convex and nonconvex functions

2 J.Dauns: A concrete approach to division rings

3 L. Butz: Connectivity in multi-factor designs. A combinatorial approach

4 P.Burmeister, B.Ganter, C.Herrmann, K.Keimel, W.Poguntke, R.Wille (eds): Universal algebra and its links with logic, algebra, combinatorics, and computer science, Proceedings of the "25. Arbeitstagung über Allgemeine Algebra", Darmstadt 1983

5 Li Weixuan: Optimal sequential block search

6 Yu.A.Kutoyants: Parameter estimation for stochastic processes, translated from the Russian and edited by B.L.S. Prakasa Rao

7 M.Jünger: Polyhedral combinatorics and the acyclic subdigraph problem

Instructions for authors are given on the rear cover.

R&E 4
research and exposition in mathematics
edited by Karl H. Hofmann and Rudolf Wille

P. Burmeister, B. Ganter, C. Herrmann
K. Keimel, W. Poguntke, R. Wille (eds.)

Universal Algebra and its links with logic, algebra, combinatorics and computer science

Proceedings of the
'25. Arbeitstagung über Allgemeine Algebra'
Darmstadt 1983

Heldermann Verlag Berlin

Deutsche Bibliothek Cataloguing in Publication Data

Universal algebra and its links with logic, algebra, combinatorics and computer science : proceedings of the "25. Arbeitstagung über Allg. Algebra", Darmstadt 1983 / ed. by P. Burmeister ... -Berlin : Heldermann, 1984.
(Research and exposition in mathematics ; Vol. 4)
ISBN 3-88538-204-0

NE: Burmeister, Peter (Hrsg.); Arbeitstagung über Allgemeine Algebra (25, 1983, Darmstadt);
GT

ISBN 3-88538-204-0

FOREWORD

More than 150 participants from 17 countries attended the 25th "Arbeitstagung über Allgemeine Algebra" which took place at the Technische Hochschule Darmstadt from the 4th to the 6th of February, 1983. This conference was one of a series of workshops on "Allgemeine Algebra" which were held since 1970 at various universities in West Germany, Austria, and Switzerland.

It was the aim of the 25th conference to discuss the state and the perspectives of research in universal algebra under the aspect of its relations to other fields of mathematics.

The discussions were stimulated by five two-hour lectures which surveyed connections to lattice theory, logic, algebra, combinatorics, and computer science. In five workshops corresponding to the main lectures, the discussions were continued and deepened; 17 invited speakers gave half-hour talks, and other participants contributed short remarks.

The present volume documents the central ideas of the main lectures and most of the other invited talks. We hope to bring to a broader public the actual orientation in universal algebra which was discussed during the conference.

We would like to thank the authors for their contributions and the referees for their careful reading of each paper. We would also like to acknowledge the financial support of the Deutsche Forschungsgemeinschaft, the Kultusminister des Landes Hessen, and the Technische Hochschule Darmstadt.

Darmstadt, December 1983

P. Burmeister, B. Ganter, C. Herrmann, K. Keimel, W. Poguntke, R. Wille

CONTENTS

Universal Algebra and its Links with Logic,
Algebra, Combinatorics, and Computer Science
Proc. "25. Arbeitstagung über Allgemeine Algebra", Darmstadt 1983
P. Burmeister et al. (eds.)

1 – 13

UNIVERSAL ALGEBRA AND LATTICE THEORY: A STORY AND THREE RESEARCH PROBLEMS

G. Grätzer

I would like to tell a story and present three research problems on universal algebra and lattice theory, with special emphasis on the interaction of these two fields. Although the topics chosen reflect my special interests, I hope they give the reader the flavour of this area of research.

The story: A free m-lattice. Lattice theory is two faceted: on the one hand, a lattice is a universal algebra $\langle L; \wedge, \vee \rangle$ with two binary operations satisfying eight identities (see, e.g. [8], § I.1); on the other hand, a lattice is a special kind of poset $\langle L; \leq \rangle$ in which any two elements have a least upper bound and greatest lower bound.

The universal algebraic approach gives lattice theory such concepts as congruence relation, free lattice, free lattice on a poset P (notation: F(P)), free product, equational class, etc.

The poset approach gives for example the concept of completeness. The two approaches are sometimes difficult to reconcile. For instance, there is no free complete lattice of 3 generators since there are complete lattices of arbitrarily large size completely generated by 3 elements.

In a series of papers [10], [11], [12], [13], [14], [15], D. Kelly and the author continued the development of the theory of m-lattices which was started by P. Crawley and R. A. Dean [2]: a poset L is an m-lattice if all subsets X with $0 < |X| < \mathfrak{m}$ have a least upper bound and greatest lower bound (see also [9]). The theory of m-lattices is an important contribution of universal algebra to lattice theory. (For a survey of some aspects of this theory, see [16].)

The development of this theory is quite lattice theoretic in nature. However, occasionally, universal algebra comes to the rescue.

Let H be the poset of Figure 1. The free $\mathfrak{m}$-lattice $D(\mathfrak{m})$ (with an additional 0 and 1) on H is shown of Figure 4. It is made up of the lattice A of Figure 2, the "mirror image" of A: the lattice B, for each i a dyadic rational, a copy C_i of the lattice $C(\mathfrak{m})$ of Figure 2 ($C(\aleph_0)$ is in the "middle" of $C(\mathfrak{m})$; the upper part of $C(\mathfrak{m})$ is not shown), and for each real t, $0 < t < 1$, t non-dyadic, a copy C_i of the two-element chain. Figure 5 shows in more detail how the elements of A and B interact with the C_i.

The result $D(\aleph_0) = F(H)$ is due to I. Rival and R. Wille [27].

The proof that $D(\mathfrak{m}) - \{Y, Y'\}$ is the free $\mathfrak{m}$-lattice on H uses a result of P. Crawley and R. A. Dean [2]; it is necessary to verify condition $(W_{\mathfrak{m}})$: if $P = \bigwedge X \leq \bigvee Y = q$, then $(X \cup Y) \cap [p, q] \neq \emptyset$. $(X, Y \subseteq D(\mathfrak{m}), 0 < |X|, |Y| < \mathfrak{m})$. The proof is very long and requires a detailed analysis of where p and q are.

In the paper [13], D. Kelly and the author found a universal algebraic proof avoiding most of the tedious computations. This proof uses the $\mathfrak{m} = \aleph_0$ case (the result of Rival and Wille) and some universal algebraic trivialities.

Let $\underset{\sim}{K}$ be a variety (equational class) of algebras of some finitary or infinitary type. For $\mathfrak{A} = \langle A; F\rangle \in \underset{\sim}{K}$ and $H \subseteq A$, we define the <u>partial</u> $\underset{\sim}{K}$<u>-algebra</u> $\mathfrak{H} = \langle H; F\rangle$ on H as follows: if $f \in F$, $a_0, a_1, \ldots \in H$ and $f(a_0, a_1, \ldots) = a \in H$ in $\mathfrak{A}$, then (and only then) $f(a_0, a_1, \ldots)$ is defined on H and equals a. $\mathfrak{H}$ is called a <u>relative algebra</u> of $\mathfrak{A}$.

If $\mathfrak{B} = \langle B; F\rangle$ is a partial algebra of the type of $\underset{\sim}{K}$, then $F(\mathfrak{B})$ denotes the free $\underset{\sim}{K}$-algebra generated by $\mathfrak{B}$. The canonical map of $\mathfrak{B}$ into $F(\mathfrak{B})$ is not necessarily one-to-one; if it is one-to-one, then it is an embedding of $\mathfrak{B}$ into $F(\mathfrak{B})$. It is an isomorphism iff $\mathfrak{B}$ is a partial $\underset{\sim}{K}$-algebra; in this case, $\mathfrak{B}$ is isomorphic to the relative algebra of $\mathfrak{A}$ on the image of B.

Now let $\mathfrak{A}_0$ and $\mathfrak{A}_1$ be partial $\underset{\sim}{K}$-algebras,

$A_0 \cap A_1 = A_2$ such that $\mathfrak{A}_2$ is the same as a relative algebra of $\mathfrak{A}_0$ and of $\mathfrak{A}_1$. We shall say that $\mathfrak{A}_0$ and $\mathfrak{A}_1$ <u>can be strongly amalgamated over</u> $\mathfrak{A}_2$, if there is an algebra $\mathfrak{A}_3 \in \underset{\sim}{K}$ of which both $\mathfrak{A}_0$ and $\mathfrak{A}_1$ are relative algebras and $A_0 \cap A_1 = A_2$ in $\mathfrak{A}_3$. The following easy lemma should give the flavour of the new proof.

<u>Lemma</u>. Let $\mathfrak{A}$ be a partial $\underset{\sim}{K}$-algebra, let $A' \subseteq A$, and let $\mathfrak{A}'$ be the corresponding relative algebra of $\mathfrak{A}$. If $\mathfrak{A}$ and $F(\mathfrak{A}')$ can be strongly amalgamated over $\mathfrak{A}'$, then the subalgebra $[A']$ of $F(\mathfrak{A})$ generated by A' is naturally isomorphic to $F(\mathfrak{A}')$.

* *

<u>First Problem</u>: Let $\mathfrak{A}$ be an idempotent algebra such that $p_i(\mathfrak{A}) < \infty$ for all i. Does there exist an n such that $\langle p_2(\mathfrak{A}), \ldots, p_n(\mathfrak{A})\rangle$ has the Minimal Extension Property?

The concept of the p_n-sequence was introduced by E. Marczewski. Take an algebra $\mathfrak{A}$, and consider the sequence $\langle f_0, f_1, f_2, \ldots\rangle$, where f_n is the size of the free algebra over $\mathfrak{A}$ with n generators. The Basic Problem is the characterization of such sequences. Now look at Figure 6, depicting the free algebra with 2 generators. The shaded areas are two copies of $F(1)$, they overlap in $F(0)$. Thus $f_2 \geq 2f_1 - f_0$. There is a similar inequality for f_n for every n (the, so called, inclusion exclusion principle). So it is more convenient to study p_n, the number of essentially n-ary polynomials ("essentially" means that they depend on all n variables). The size of the unshaded area in Figure 6 is p_2.

So let $p_0(\mathfrak{A})$ be the number of constant unary polynomials, let $p_1(\mathfrak{A})$ be the number of non-constant (essentially unary) polynomials excluding $p(x) = x$; for $n > 1$, $p_n(\mathfrak{A})$ is the number of essentially n-ary polynomials.

The Basic Problem restated is: which sequences $\langle p_n\rangle$ can be represented as $\langle p_n(\mathfrak{A})\rangle$ for some algebra $\mathfrak{A}$. It was proved in [22] that if $p_0 \neq 0$, $\langle p_n\rangle$ is always so representable.

In the case $p_0 = 0$, there are many representation

results. The following, also from [22], is typical: If $p_n > 0$ for all $n \geq 1$, or if n divides p_n for n even and $p_n > 0$ for n odd, or if n divides p_n for all n, then $\langle p_n \rangle$ is so representable.

All three conditions impose some restriction on each p_n, but the size of p_i does not influence the size of p_j if $i \neq j$.

In the idempotent case ($p_0 = p_1 = 0$, that is, there are no nullary operations and $f(x, \ldots, x) = x$ for all operations f) the situation is very different ([21] and [24]):

Let $\mathfrak{A}$ be an idempotent algebra which is not equivalent to a semilattice, a diagonal algebra ($\langle B^n; f \rangle$, where $f(\langle a_1^1, \ldots, a_n^1 \rangle, \langle a_1^2, \ldots, a_n^2 \rangle, \ldots, \langle a_1^n, \ldots, a_n^n \rangle) = \langle a_1^1, a_2^2, \ldots, a_n^n \rangle$ defines the n-ary operation f), an idempotent reduct of a Boolean group (let $\langle G; + \rangle$ be a Boolean group, that is, $2x = 0$; the idempotent reduct is $\langle G; x + y + z \rangle$). Then there exists an n such that

$$p_n(\mathfrak{A}) < p_{n+1}(\mathfrak{A}) < \ldots$$

So in the idempotent case $\langle p_n \rangle$ is strictly increasing from some point on (with three exceptions). Of course, given a $\langle p_i(\mathfrak{A}) \rangle$, we can trivially construct a $\langle p_i(\mathfrak{B}) \rangle$ which agrees with $\langle p_i(\mathfrak{A}) \rangle$ up to n, and from n on $\langle p_i(\mathfrak{B}) \rangle$ increases faster: $p_i(\mathfrak{B}) > p_i(\mathfrak{A})$ for all $i > n$.

For the free idempotent semigroup satisfying $xyz = xzy$, obviously $p_n = n$ for all $n \geq 2$. The following result of J. Płonka [26] proves an astonishing converse: Let $\mathfrak{A}$ be an idempotent algebra satisfying $p_2(\mathfrak{A}) = 2$, $p_3(\mathfrak{A}) = 3$, and $p_4(\mathfrak{A}) = 4$. Then $p_n(\mathfrak{A}) \geq n$ for all $n \geq 2$.

In other words, $\langle 0, 0, 2, 3, 4 \rangle$ has a "minimal" extension $\langle 0, 0, 2, 3, 4, 5, \ldots, n, \ldots \rangle$; every other extension is $\geq$ the minimal extension.

Definition: $\langle 0, 0, p_2, \ldots, p_n \rangle$ has the Minimal Extension Property iff there is an idempotent algebra $\mathfrak{A}$ satisfying (i) $p_i = p_i(\mathfrak{A})$ for $2 \leq i \leq n$; (ii) for any idempotent algebra $\mathfrak{B}$, if $p_i = p_i(\mathfrak{B})$ for $2 \leq i \leq n$, then $p_j(\mathfrak{A}) \leq p_j(\mathfrak{B})$ for all $j \geq n$.

It was proved in my paper with R. Padmanabhan [17] that $\langle 0, 0, 1, 3, 5 \rangle$ has the Minimal Extension Property and the minimal extension is given by the algebra $\langle A; \circ \rangle$, where $\langle A; + \rangle$ is an abelian group of exponent 3, and $x \circ y = 2x + 2y$.

The First Problem asks, whether any idempotent algebra $\mathfrak{A}$ gives rise to a sequence $\langle 0, 0, p_i(\mathfrak{A}), \ldots, p_n(\mathfrak{A}) \rangle$ with the Minimal Extension Property if n is large enough.

For a more detailed introduction to this problem, see [4]. For more recent references, consult Algebra Universalis, especially, A. Kisielewicz [24].

* *

Second Problem: Find all lattices K with the property that whenever K can be embedded in the ideal lattice $I(L)$ of a lattice L, then K can also be embedded into L.

This problem illustrates how developments in universal algebra open up new fields in lattice theory. Chapter 7 of [7] is devoted to the following question in universal algebra: which first ordered properties are preserved under algebraic constructions (e.g., formations of subalgebras, homorphic images, direct products).

The formation of the ideal lattice $I(L)$ of a lattice L is an algebraic construction that is special to lattice theory. So it is natural to raise the question what properties are preserved under the formation of ideal lattices.

I discussed this question briefly in 1961 (see [5]). Since there seem to be too many such properties, I proposed to investigate the special case: the lattice has a sublattice isomorphic to a given lattice K. This is how we arrive at the Second Problem.

Lattices K satisfying this property are called *transferable*. There is also a stronger concept. Let φ be an embedding of K into $I(L)$; for every $a \in K$, $a\varphi$ is an ideal of L. If K is transferable, then there is an embedding ψ of K into L. Now it seems natural to require, for $a \in K$, that $a\psi$ be in the ideal $a\varphi$, but not in any $b\varphi$ where $b < a$. If there is always such a map ψ, K is called *sharply transferable*.

It is an instructive exercise to check that N_5 is sharply transferable. (In fact, K. A. Baker and A. Hales observed that all finite projective lattices are transferable; Algebra Universalis 4 (1974), 250-258.) Sharp transferability is easier to handle than transferability since we know roughly where $a\psi$ should be. Let us start the discussion with some definitions:

1. Let $\langle P; \leq \rangle$ be a partially ordered set. For $X, Y \subseteq P$, define $X < Y$ to hold if and only if for every $x \in X$ there exists $y \in Y$ such that $x \leq y$.

2. Let $\langle S; \vee \rangle$ be a join-semilattice, $p \in S$, and $J \subseteq S$. We say that $\langle p, J \rangle$ is a minimal pair if and only if the following three conditions hold:

(i) $p \notin J$;

(ii) $p \leq \bigvee$;

(iii) if $J' \subseteq S$, $p \leq \bigvee J'$, and $J' < J$, then $J \subseteq J'$

3. A semilattice $\langle S; \vee \rangle$ is said to satisfy (T) if and only if there exists a linear order relation R on S such that if $\langle p, J \rangle$ is a minimal pair, then $p \, R \, x$ holds for all $x \in J$.

4. A lattice $\langle L; \wedge, \vee \rangle$ is said to satisfy $(T_\vee)$ if and only if $\langle L; \vee \rangle$ satisfies (T), and to satisfy $(T_\wedge)$ if and only if the dual of $\langle L; \wedge \rangle$ satisfies (T).

<u>Theorem</u>. A finite lattice is sharply transferable if and only if it satisfies the three conditions $(T_\vee)$, $(T_\wedge)$, and (W).

Condition (W) is (W_m) of "The story" with $m = \aleph_0$, that is, the condition: $x \wedge y \leq u \vee v$ implies that $\{x, y, u, v\} \cap [x \wedge y, u \vee v] \neq \emptyset$.

Later, in [19], I generalized this result with Platt to arbitrary lattices.

The first result for transferability is in [5] (see [6] for a proof): A transferable lattice cannot have doubly reducible elements.

The proof relies on the following result: Every lattice can be embedded in the ideal lattice of a lattice without doubly irreducible elements.

This led to the following

Conjecture. The class of transferable lattices is the intersection of all classes $\underset{\sim}{K}$ of lattices with the property that every lattice can be embedded in the ideal lattice of some lattice in $\underset{\sim}{K}$.

Such classes were investigated in [18] and [20]. Sufficiently many such classes were constructed to conclude:

Theorem. For finite lattices, transferability and sharp transferability are equivalent.

It is quite likely that proving the Conjecture would lead to a solution of the Second Problem.

* *

Third Problem: Find a nontrivial class $\underset{\sim}{K}$ of groupoids (algebras with one binary operation $+$) such that for every $\mathfrak{A} \in K$, $a, b, c, d \in A$, $c \equiv d\ (\Theta(a, b))$ iff $c = a + y$ and $d = b + y$ for some $y \in A$.

This problem comes from a universal algebraic problem via lattice theory.

For an algebra $\mathfrak{A}$, $a, b \in A$, let $\Theta(a, b)$ denote the smallest congruence relation under which $a \equiv b$. Mal'cev's Lemma (see, e.g., [7], Theorem 10.3) describes $\Theta(a, b)$ as follows:

$c \equiv d\ (\Theta(a, b))$ iff there exists a sequence $z_0 = c$, $z_1, \ldots, z_n = d$ of elements of A such that for each $i < n$, there exists a polynomial $p_i(x, x_1, \ldots, x_m)$ with $p_i(a, y_1, \ldots, y_m) = z_{i+f(i)}$, $p_i(b, y_1, \ldots, y_m) = z_{i+1-f(i)}$ for some $y_1, \ldots, y_m \in A$, where f is a choice function, $f(i) = 0$ or $f(i) = 1$.

The choice of n, of $p_0, \ldots, p_{n-1}$, and of f depends on a, b, c, and d. Would it not be nice, if they could be chosen independently of a, b, c, d? In most classical algebraic systems this is not the case. Take a Boolean group: $c \equiv d\ (\Theta(a, b))$ iff $c + d = a + b$ or $c + d = 0$; in this case, $n = 1$, $p(x, y) = x + y$, but f depends on a, b, c, and d.

For an interesting example, lattice theory came to the rescue. I proved with E. T. Schmidt [23] that in a distributive lattice L, $a, b, c, d \in L$, $a \leq b$, $c \leq d$, $c \equiv d\ (\Theta(a, b))$ iff $(a \vee y_1) \wedge y_2 = c$, $(b \vee y_1) \wedge y_2 = d$ for some $y_1, y_2 \in L$. Thus there are examples where $p_0, \ldots, p_{n-1}, f$ are the same for a whole equational class.

$p_0, \ldots, p_{n-1}$, and f is called a <u>congruence scheme</u> Σ. An equational class $\underset{\sim}{K}$ is said to have a <u>uniform congruence scheme</u>, Σ, if for all $\mathfrak{A} \in K$, $a, b, c, d \in A$, $c \equiv d\ (\Theta(a, b))$ can always be described with the same $p_0, \ldots, p_{n-1}$, and f. This concept was introduced by R. Magari [25]. See also [3].

In [1], I considered with J. Berman the question how congruence schemes may look like. If there are constants, the problem seems very difficult. Even $n = 1$, $f(0) = 0$, $p(x, y) = 0 + ((0 + x) + y)$ cannot be the congruence scheme for an algebra with more than one element. Now if there are no constant operations, then the result is very nice:

<u>Theorem</u>. $p_0, \ldots, p_{n-1}, f$ is the uniform congruence scheme for a nontrivial variety iff all p_i are at least binary.

The condition, all p_i are binary, simply means that no p_i has only x as variable.

The equational class constructed in the proof is of unspecified type. Of course, the type contains all the operations needed to build up the p_i, but it contains a number of additional operations. The Third Problem asks, in the simplest possible case, what happens if the type is fixed.

The congruence scheme in the Third Problem is $p(x, y) = x + y$ (and $f(0) = 0$). One can raise the same problem with any other reasonable polynomial $(y + (x + y),\ y + ((x + y) + x)$, etc.) or pairs of polynomials, e.g., $p_0(x, y) = x + y$, $p_1(x, y) = y + ((x + y) + (x + y))$ $f(0) = 0$, $f(1) = 1$ that is,

$$c \equiv d\ (\Theta(a, b))$$

iff

$$c = a + y_1,\ b + y_1 = y_2 + ((a + y_2) + (a + y_2)),$$
$$d = y_2 + ((b + y_2) + (b + y_2)).$$

References

[1] J. Berman and G. Grätzer: Uniform representations of congruence schemes. Pacific J. Math. 76 (1978), 301-311.

[2] P. Crawley and R. A. Dean: Free lattices with infinite operations. Trans. Amer. Math. Soc. 92 (1959), 35-47.

[3] E. Fried, G. Grätzer, and R. W. Quackenbush: Uniform congruence schemes. Algebra Universalis 10 (1980), 176-188.

[4] G. Grätzer: Composition of functions. In Proceedings of the Conference on Universal Algebra, Queen's University, Kingston, Ont. (1969), 1-106.

[5] G. Grätzer: Universal Algebra. In Current Trends in Lattice Theory. D. Van Nostrand, (1970), 173-215.

[6] G. Grätzer: A property of transferable lattices. Proc. Amer. Math. Soc. 43 (1974), 269-271.

[7] G. Grätzer: Universal Algebra. Second Edition. Springer Verlag, New York; Heidelberg, Berlin, 1979.

[8] G. Grätzer: General Lattice Theory. Pure and Applied Mathematics Series, Academic Press, New York, N. Y.; Mathematische Reihe, Band 52, Birkhäuser Verlag, Basel; Akademie Verlag, Berlin, 1978. (Russian translation: MIR Publishers, Moscow, 1982.)

[9] G. Grätzer, A. Hajnal, and D. Kelly: Chain conditions in free products of lattices with infinitary operations. Pacific J. Math. 83 (1979), 107-115.

[10] G. Grätzer and D. Kelly: A normal form theorem for lattices completely generated by a subset. Proc. Amer. Math. Soc. 67 (1977), 215-218.

[11] G. Grätzer and D. Kelly: Free $\mathfrak{m}$-products of lattices. I and II. Colloq. Math., to appear.

[12] G. Grätzer and D. Kelly: The free $\mathfrak{m}$-lattice on the poset H. ORDER, to appear.

[13] G. Grätzer and D. Kelly: A technique to generate $\mathfrak{m}$-ary free lattices from finitary ones.

[14] G. Grätzer and D. Kelly: An embedding theorem for free $\mathfrak{m}$-lattices on slender posets.

[15] G. Grätzer and D. Kelly: A description of free $\mathfrak{m}$-lattices on slender posets.

[16] G. Grätzer and D. Kelly: The construction of some free m-lattices on posets. Proceedings of the Lyon Conference on Ordered Sets.

[17] G. Grätzer and R. Padmanabhan: On idempotent, commutative, and non-associative groupoids. Proc. Amer. Math. Soc. 28 (1971), 75-80.

[18] G. Grätzer and C. R. Platt: Two embedding theorems for lattices. Proc. Amer. Math. Soc. 69 (1978), 21-24.

[19] G. Grätzer and C. R. Platt: A characterization of sharply transferable lattices. Canad. J. Math. 32 (1980), 145-154.

[20] G. Grätzer, C. R. Platt, and B. Sands: Embedding lattices into lattice of ideals. Pacific J. Math. 85 (1979), 65-75.

[21] G. Grätzer and J. Plonka: On the number of polynomials of an idempotent algebra. I and II. Pacific J. Math. 32 (1970), 697-709 and 47 (1973), 99-113.

[22] G. Grätzer, J. Plonka, and A. Sekanina: On the number of polynomials of a universal algebra. I. Colloq. Math. 22 (1970), 9-11.

[23] G. Grätzer and E. T. Schmidt: Ideals and congruence relations in lattices. Acta Math. Acad. Sci. Hungar. 9 (1958), 137-175.

[24] A. Kisielewicz: The p_n-sequences of idempotent algebras are strictly increasing. Algebra Universalis 13 (1981), 233-250.

[25] R. Magari: The classification of idealizable varieties (Congruenze ideali IV), J. of Alg., 26 (1973), 152-165.

[26] J. Płonka: On algebras with at most n distinct essentially n-ary operations. Algebra Universalis 1 (1971), 80-85.

[27] I. Rival and R. Wille: Lattices freely generated by partially ordered sets: which can be "drawn"?. J. Reine. Angew. Math. 310 (1979), 56-80.

University of Manitoba
Winnipeg, Manitoba
R3T 2N2 Canada

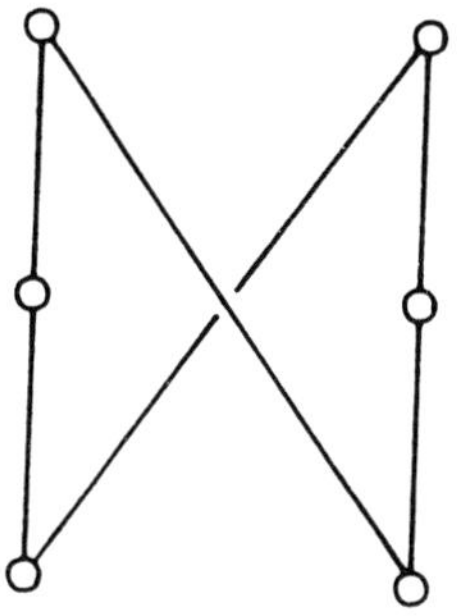

The poset H

Figure 1

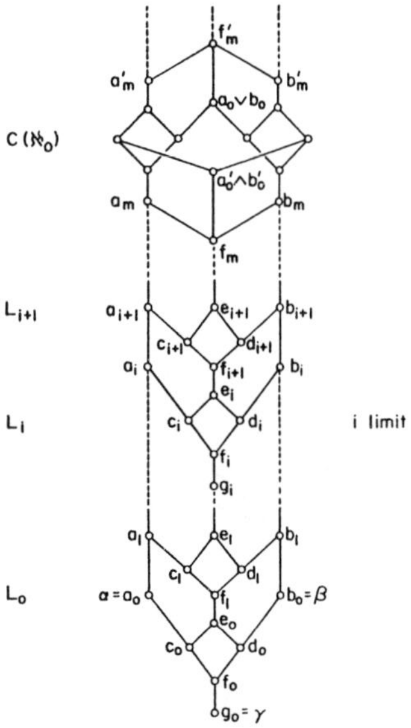

The m-lattice C(m)

Figure 2

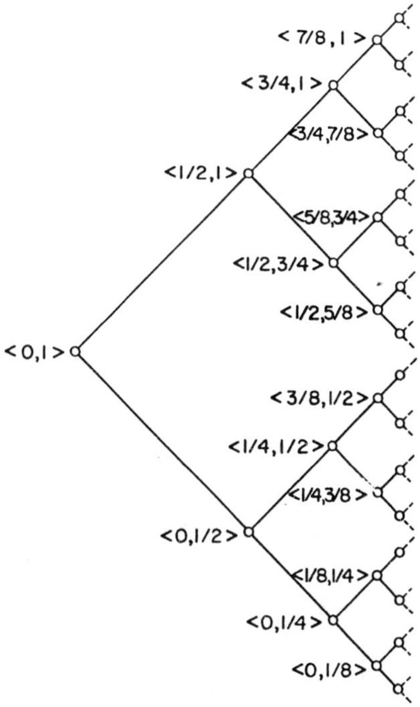

The lattice A

Figure 3

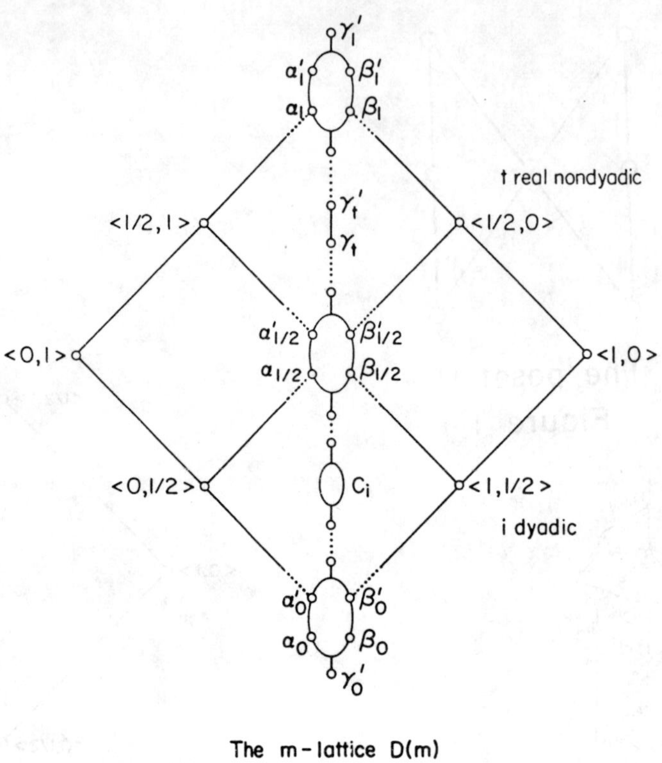

Figure 4

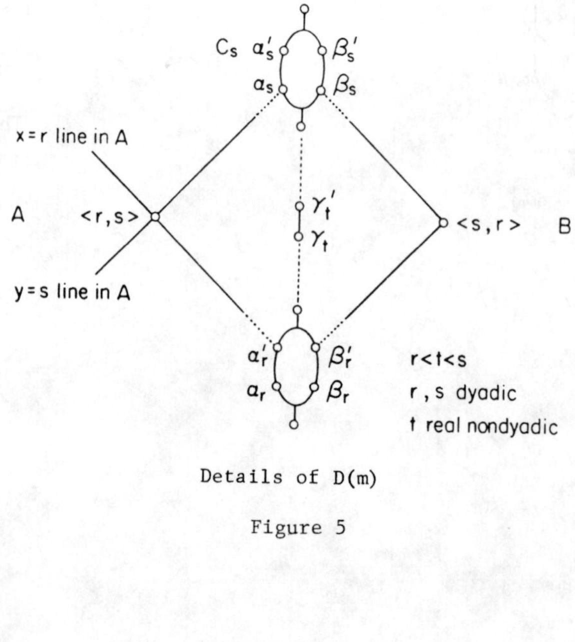

Details of D(m)

Figure 5

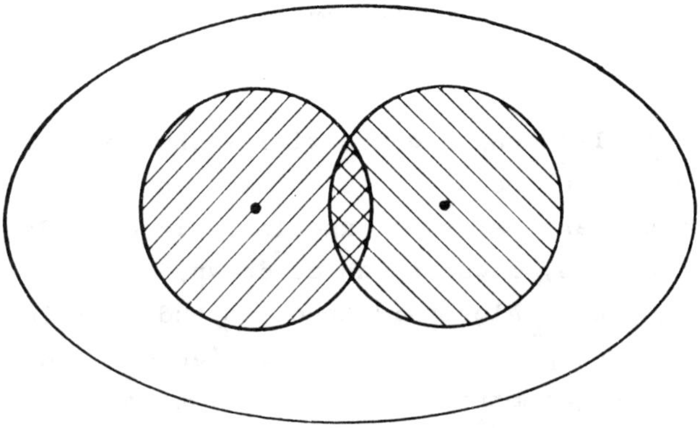

Figure 6

Universal Algebra and its Links with Logic,
Algebra, Combinatorics, and Computer Science
Proc. "25. Arbeitstagung über Allgemeine Algebra", Darmstadt 1983
P. Burmeister et al. (eds.)

14 – 28

GEOMETRICAL REASONING AND ANALOGY IN UNIVERSAL ALGEBRA

H. P. Gumm

Most students of classical algebra will sooner or later appreciate the close connection between algebraic and geometric concepts. Geometric concepts are parts of experience in everyday life. They are easy to visualize, figures can be drawn, rules and theorems can actually be seen. Algebraic theorems and formulas are usually much harder to grasp for the unexperienced. Many textbooks therefore dwell in geometrical visualizations and proofs for some simple algebraic theorems like the binomial or the Pythagorean theorem. It takes years of apprenticeship until a familiarity with the permitted manipulations of abstract strings and formulas is achieved. Geometric visualization is not restricted to the segment of geometry that is actually being experienced. It is easy to invent analogies. As an example consider the classification of the possible sets of solutions to a system of linear equations. Even though a many - dimensional space is not a part of experience it is easy to imagine the different hyperplanes that intersect to give infinitely many, exactly one, or no solution. And in devising a proof it is very helpful to keep the geometric pictures in mind to stake out the rough lines for a proof. Later the gaps will have to be filled with calculations.

What is then the geometry, what are the geometric pictures that can aid in studying universal algebras? The experience with classical algebra first gives us the choice between a projective approach and an affine approach. What should the projective geometry associated with a universal algebra be? In the classical case of modules there is a one-to-one correspondence between subspaces and the congruence relations they induce. So in the universal algebra case, the subspaces of the projective (pseudo-) geometry should correspond to the congruence relations of the algebra A ; thus studying the lattice Con(A) of congruences on A would be studying the projective geometry of A. Geometrically important concepts as e.g. the Desarguesian or the Pappian laws can be translated into lattice theoretical concepts for Con(A). A first theorem in this environment is due to B.Jonsson /17/:

Let A have permutable congruences, then Con(A) is arguesian.

Here the arguesian law is nothing but the lattice theoretical phrasing of the Desarguesian law, thus the conclusion of the theorem is:

The projective geometry of A is Desarguesian.

The hypothesis requires that for any two congruences Θ and Ψ, the relational product $\Theta\circ\Psi$ is a congruence again, which is the same as saying $\Theta\circ\Psi = \Psi\circ\Theta$. This condition is well known to imply that Con(A) is a modular lattice, and indeed R.Freese and B.Jonsson were able to show in / 6/:

Let A and all subdirect subalgebras of $A\times A$ have modular congruence lattices. Then Con(A) is arguesian.

Here it is not enough if A has a modular congruence lattice. Freese and Jonsson use a simple trick which is both easy and well known but very helpful in what follows:

Trick: Every congruence relation Θ on an algebra A is a subalgebra of $A\times A$. Θ is actually embedded subdirectly in $A\times A$.

It is exactly those subdirect powers of A which Freese and Jonsson require to have modular congruence lattices.

Other "projective" investigations concern finite algebras whose projective geometry is one-dimensional. It had for a long time been open whether the number of nontrivial subspaces would always have to be one more than a prime power. A counterexample was given by W.Feit / 5 /, resting on work of P.Palfy and P.Pudlak / 19/.

In the affine approach which was introduced and systematically studied by R.Wille /21/, we let the points correspond to elements of the algebra A, and we take as lines the classes of congruence relations on A. A class of a congruence relation Θ will be called a Θ-line. The collection of classes of a fixed congruence

relation then will be a class of pairwise parallel lines.
This enables us to draw pictures where points represent elements of A and lines represent congruence classes of some congruence Θ, in which case we sometimes lable the line with the letter Θ and call them "Θ-lines". We shall draw two lines parallel if they represent classes of the same congruence relation.
As a first example of the connections between algebraic and geometric properties we look at permuting congruence relations Θ and Ψ. The fact that Θ and Ψ permute can geometrically be expressed as follows:
Let l be a Θ-line and g a Ψ-line and suppose l and g intersect with y a point of intersection. For any points x on l and z on g there exists a point u making (x,y,z,u) a parallelogram. Pictorially:

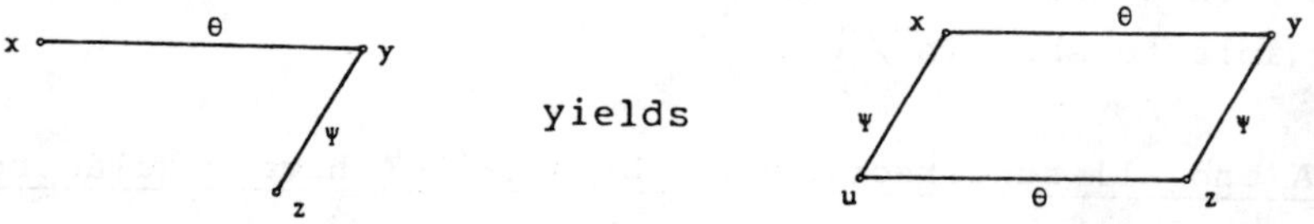

The case that all congruences in all algebras of some variety permute is captured in Mal'cevs famous theorem, stating that this is equivalent to the existence of a ternary term $m(x,y,z)$ satisfying the equations $m(x,y,y) = x$ and $m(x,x,y) = y$. Since m is compatible with all congruences, it is easy to see that this is in turn equivalent to saying that $m(x,y,z)$ is always a fourth parallelogram point in the situation of the above figure.
The equations are abtained by letting Θ or Ψ shrink to the identity congruence. This simple <u>parallelogram principle</u> was extremely fruitful for further investigations. So it was easy to conclude that the congruence class geometries of algebras in such a "permutable" variety obey the <u>Little law of Desargues</u> which is given pictorially as

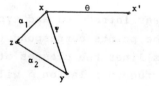

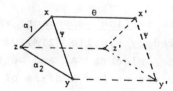

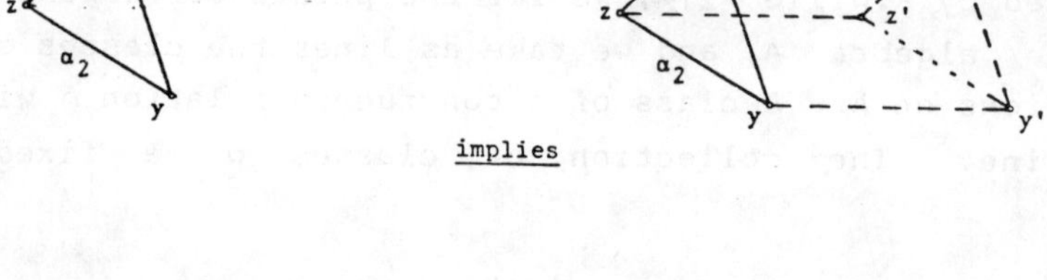

This is to be read as: Given x,y,z,x', connected by ψ, α_1, α_2 and Θ-lines there exist y' and z' with the relations indicated, where lines drawn parallel in the figure mean lines of the same congruence relation. Here y' and z' can be constructed as m(x',x,y) resp. m(x',x,z).
If we let x and y coincide, i.e. we take $\Psi = 0$, the identity congruence, then we see that the parallelogram principle follows. In other words, the varieties in which the Little Desarguesian law holds are precisely the permutable varieties. Unfortunately this result seems not to fit in with the projective version of Freese and Jonsson that, projectively, the Desarguesian theorem characterizes the modular varieties. The resolution of this problem though will be achieved when we come to slightly reformulating the little Desarguesian law.

Affine geometry in modular algebras

To do affine geometry in congruence modular algebras we need some geometrical substitute for the parallelogram principle. This was found in the following principle /8 /:

Shifting Lemma: Let α, β and γ be congruences in a congruence modular algebra A and x,y,z,u be points of A. Suppose moreover that $\alpha \wedge \beta \leq \gamma$ then

implies

The proof of this lemma is extremely simple, nevertheless the lemma itself is probably the most important tool for the geometrical study of modular varieties. The second tool we need is the simple trick we have already met, to think of a congruence Θ on A as a subalgebra of A×A.
We shall show now how these two ingredients, or simple modifications thereof suffice to develop the affine geometry of congruence modular algebras.

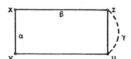

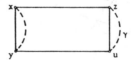

Let us start with the theorems of Desargues and Pappus. Clearly we must modify the formulation of Desargues' law slightly to /9/:
<u>The Little Desarguesian law:</u> Let $\Theta,\alpha_1,\alpha_2,\Psi$ be congruences with $\Theta\wedge\alpha_1 \leq \Psi$ and $\Theta\wedge\alpha_2 \leq \Psi$ and let x,y,z,x',y',z' be elements of A then

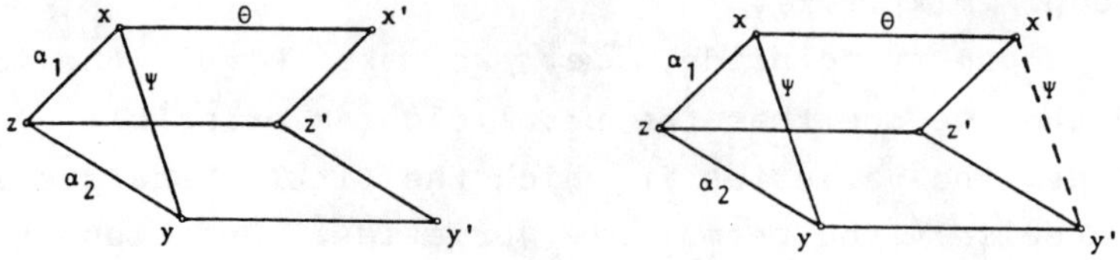

<u>implies</u>

We note that as in the shifting lemma we require all necessary points already to exist and we need "dimensionality conditions" for the congruences (resp. subspaces) involved. We obtain, in complete analogy to the Freese - Jonsson result:

<u>Let A and all subdirect subalgebras of A×A have modular congruences. Then the affine geometry of A satisfies the Little Desarguesian law.</u>

The proof becomes more transparent if we restate the theorem in a slightly more general version, namely:
Given points $x,y,z,u,x',y'z',u'$ and congruences $\Theta,\alpha_1,\alpha_2,\Psi$ with $\Theta\wedge\alpha_1 \leq\Psi$ and $\Theta\wedge\alpha_2\leq \Psi$ then

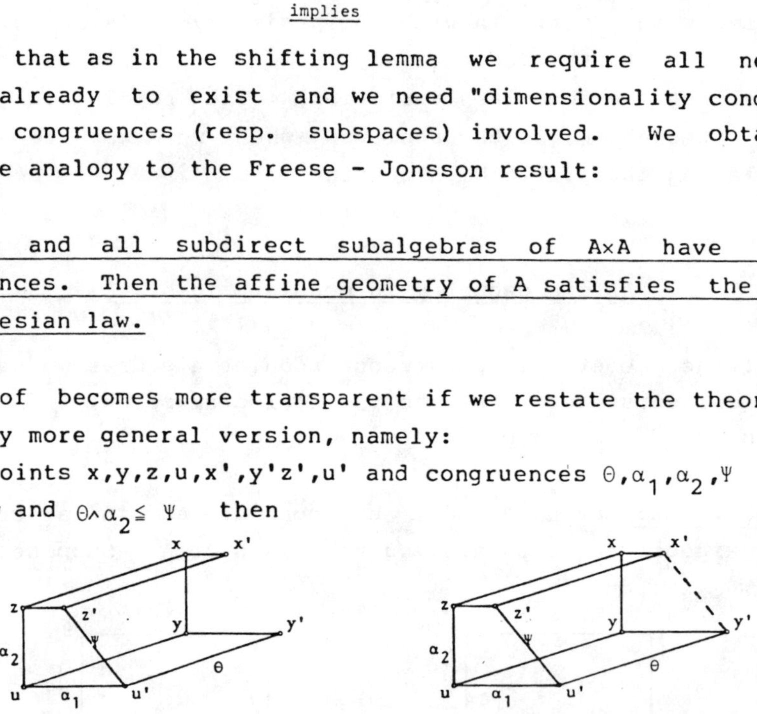

<u>implies</u>

The idea for a proof develops when we imagine this figure as a three-dimensional configuration and look at it "from the side", namely along the α_1 -lines. The figure we see is

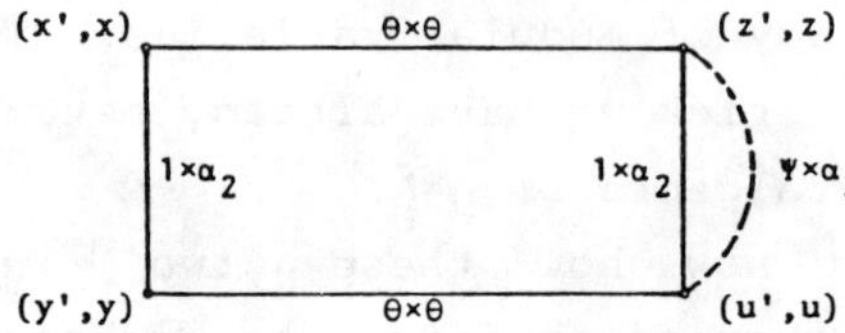

where now the points are elements of <u>the algebra α_1</u>. If we can apply the shifting lemma to this situation, showing that (x',x) $\Psi\times\Theta$ (y',y) we are done. Well, by hypothesis the algebra α_1 again is congruence modular, thus we are left to check that in this algebra $1\times\Theta \wedge \Theta\times\alpha_2 \leq \Psi\times\Theta$. This is true for elements (a,b), (c,d) from the algebra α_1 by a second application of the shifting lemma. For ((a,b),(c,d)) from the left hand side we obtain, using the fact that $\Theta\wedge\alpha_2 \leq \Psi$, the configuration

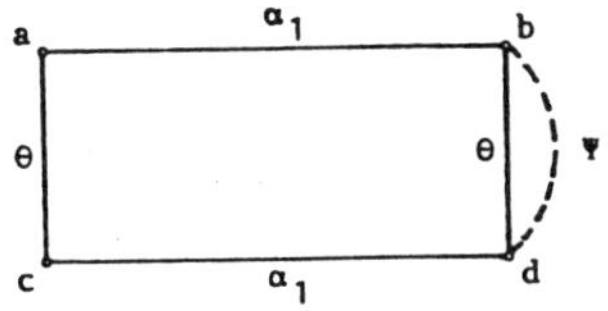

from which by the shifting lemma we may conclude $(a,c) \in \Psi$, so $((a,b),(c,d)) \in \Psi\times\Theta$.

Analogously we may formulate a version of the <u>Little Pappian theorem</u>:

Let $\Theta, \Psi, \alpha_1, \alpha_2$ be congruences with $\Theta\wedge\alpha_1 \leq \Psi$ and $\Theta\wedge\alpha_2 \leq \Psi$ and x,y,z,u,x',u' points then we obtain with a similar proof:

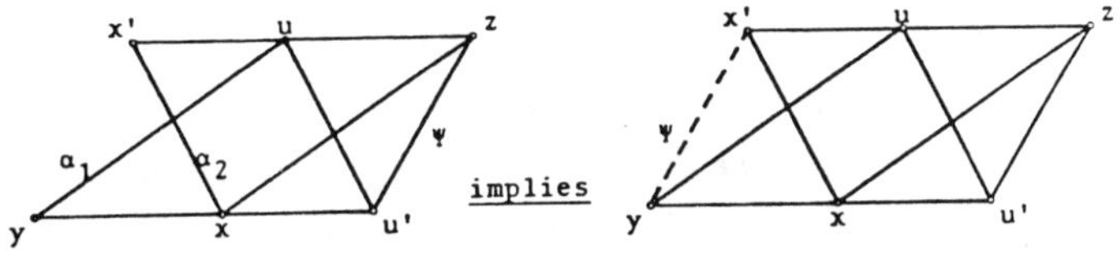

<u>Let A and all congruences of A have modular congruence lattices. Then the affine geometry of A is Pappian.</u>

At this point some interesting questions may be raised: Since the analogy of projective and affine versions seems perfect in the case of the Desarguesian law there should be a similar analogy with the Pappian law. Thus a projective version of the Pappian law should be discovered for modular varieties. More generally an interesting question arises: <u>Is there a general translation between affine and projective properties of congruence modular algebras?</u>

Reviewing some known theorems

Once certain principles and ways of reasoning have been applied successfully one may review known theorems trying to look at them in a new light. We elaborate two examples. Before we go on though we shall say a few more words about the relationship of congruence modularity and the shifting lemma. On the level of varieties the shifting lemma is equivalent to modularity. On the level of single algebras though, modularity lies between the shifting lemma and a slight generalization thereof, the shifting principle /10/:

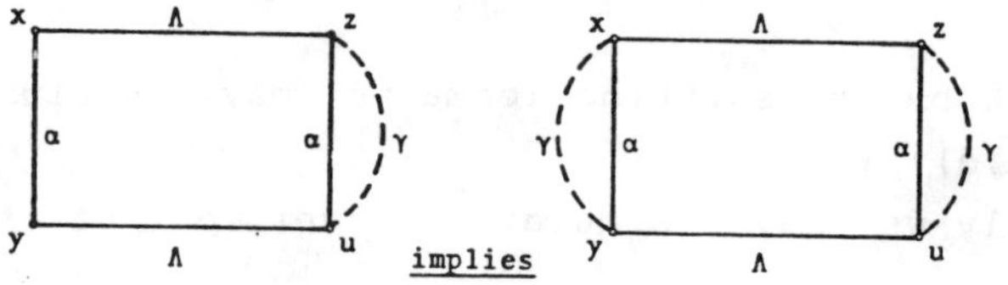

where α and γ are congruences, but Λ may be any reflexive, symmetrical subalgebra of $A \times A$. As in the shifting lemma we assume $\alpha \wedge \Lambda \leq \gamma$. W.l.o.g. $\gamma \leq \alpha$, otherwise replace γ by $\gamma \wedge \alpha$. Thus to show modularity, it is enough to show the shifting principle. We now come to our first example, which is an improvement of a theorem of J.Hagemann /14/, due to S.Bulman-Fleming, A.Day and W.Taylor/ 2 /:

Let all subalgebras of $A \times A$ have regular congruences, then A has modular congruences.

Here A is said to have regular congruences, if congruences are uniquely determined by any of their classes.

To prove the shifting principle, we look at Λ, the subalgebra of $A \times A$, and need to show that in this algebra we have $\gamma \times \alpha = \gamma \times \gamma$. Since the above version of the shifting principle becomes trivial for the case a = c:

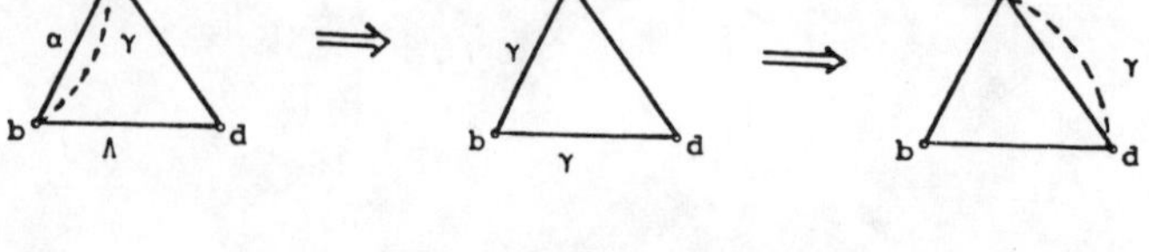

it follows that the (a,a)-class of $\gamma\times\alpha$ coincides with the (a,a)-class of $\gamma\times\gamma$, thus $\gamma\times\alpha = \gamma\times\gamma$ by regularity.

As a second example we consider refinement theorems for direct products. If $A\times B \cong C\times D$ one wants to have canonical refinements (see /1/) such that $A \cong E_1\times E_2$, $B \cong F_1\times F_2$, $C \cong E_1\times F_1$ and $D \cong E_2\times F_2$. Let $A\times B$ be given with the canonical projection kernels π_1 and π_2 and consider the congruences Θ and Ψ on $A\times B$ yielding the decomposition corresponding to $C\times D$. Then the shifting lemma yields that a congruence relation Θ_B may be defined on B as

$$b_1\, \Theta_B\, b_2 \Leftrightarrow \exists x \quad (x,b_1)\ \Theta\ (x,b_2)$$
$$\Leftrightarrow \forall y \quad (y,b_1)\ \Theta\ (y,b_2)$$

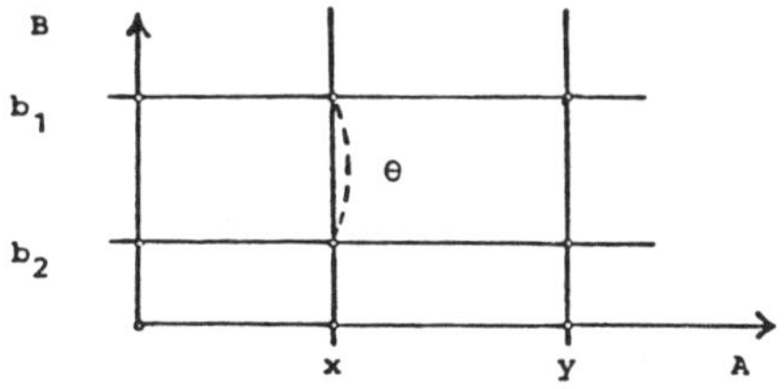

One has canonical refinements if and only if in such a way the congruences Θ and Ψ split in congruence relations Θ_A, Θ_B, Ψ_A and Ψ_B, which provide the factors for the refinement.
Here the shifting lemma is being applied in a very special situation and the argument is not restricted to congruence modular algebras. Even the fact that Θ and Ψ split in the desired way may be formulated as an easy geometrical principle. In this spirit H.Bauer and R.Wille /1/ have given an elegant proof for Hashimoto's refinement theorem for products of posets (/16/).
In modular varieties Θ and Ψ do generally not split as requested. So called "skew" congruences have to be introduced, but they too can be geometrically analysed so that for modular varieties and in the presence of chain conditions one obtains a cancellation and refinement theorem "up to isotopy". In the proof of this result /13/, unfortunately affine and projective reasoning is being mixed so a "purer" proof of this result using the above methods would be desirable.

Using geometry to guide syntactical inferences

The difference between the configuration theorems encountered so far in permutable and in modular varieties can shortly be subsumed by:

In permutable varieties points with the desired relations may be generated, whereas in modular varieties the points already are supposed to exist satisfying some of the desired relations and the rest of the relations can be concluded.

The prime example for this distinction is the comparison of the parallelogram principle versus the shifting lemma.

We shall see in this chapter though, that we do have a version of the parallelogram principle in modular varieties. The parallelogram principle holds, provided some auxiliary points are given. Moreover, the fourth-parallelogram point is provided by a ternary term t(x,y,z) which does share many properties with a Mal'cev term / 8 /:

In every modular variety there exists a ternary term t(x,y,z) such that t(x,y,y) = x, and, given a configuration

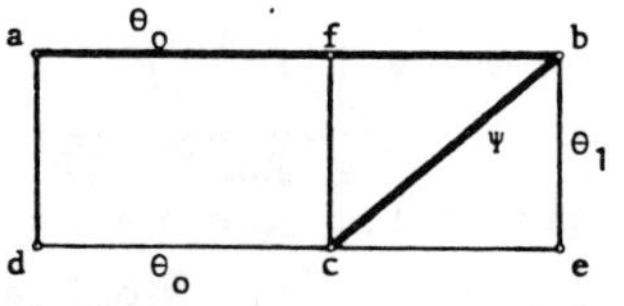

with $\Theta_0 \wedge \Theta_1 \leq \Psi$, t(a,b,c) completes a,b,c to a Θ_0-Ψ-parallelogram.

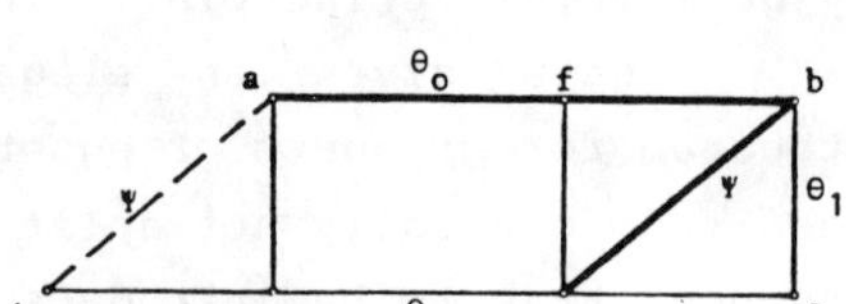

The proof of this theorem is a prime example of how geometrical analogy may be used to stake out a proof and fill the gaps later with calculations. We shall give this geometric idea. It is noteworthy that the proof implicitly involves multiple substitutions of terms into other terms.

We start with the terms and equations given by A.Day /4 /, describing congruence modularity: A variety $\underline{V}$ is congruence modular iff there is a number n and quaternary terms $m_0, m_1, \ldots, m_n$ such that the following equations are true in $\underline{V}$:

(M0) $m_0(x,y,z,u) = x$

(M1) $m_i(x,x,y,y) = x$ for all $0 \leqslant i \leqslant n$,

(M2) $m_i(x,y,x,y) = m_{i+1}(x,y,x,y)$ for $0 \leqslant i < n$, i even

(M3) $m_i(x,y,z,z) = m_{i+1}(x,y,z,z)$ for $0 \leqslant i < n$, i odd

(M4) $m_n(x,y,z,u) = y$.

Let us draw the result of applying these terms to the points in the given configuration. We define $\underline{m}_i := m_i(a,d,b,c)$. Then the equations give us relations between these points $\underline{m}_i$. Using (M0), (M2), (M3), (M4), we obtain a figure which we draw for the case n = 7:

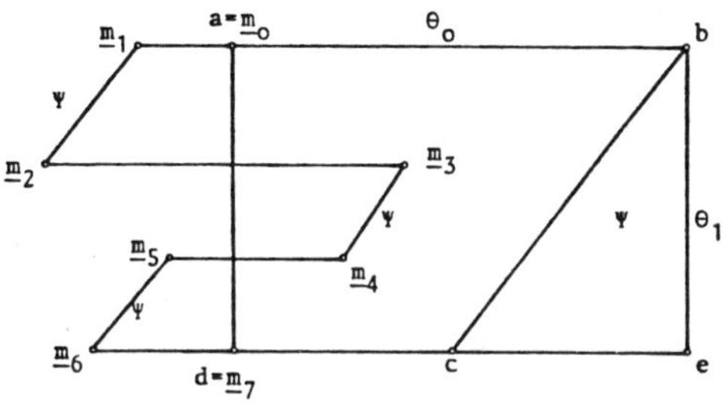

Now the rough idea how to construct a Ψ-line from a to the bottom line becomes immediately apparent: The little Ψ-pieces between $\underline{m}_i$ and $\underline{m}_{i+1}$ for odd i should be shifted to the appropriate positions, starting at a, to connect to the desired line. Substituting a and d in the first places of the Day-terms and substituting some of b,c,e in the last two places, more precisely, setting

$\underline{t}_i := m_i(a,d,c,e)$ for i odd, $\underline{t}_i := m_i(a,d,b,c)$ for i even,

$\underline{s}_i := m_i(a,d,c,c)$ for i odd, $\underline{s}_i := m_i(a,d,b,e)$ for i even

we obtain a configuration as in

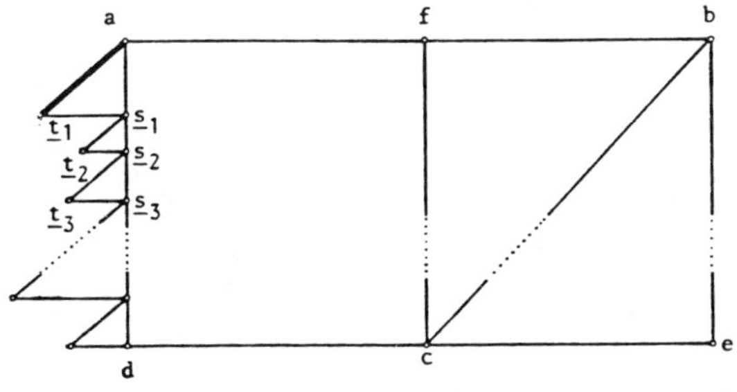

Now we do have some control about the horizontal position of the little Ψ-pieces. Consider t_1, the first of the resulting peaks, then it is easy to find a point a^1 on the top line and d^1 on the bottom line above , resp. below t_1. Applying the terms as before, but replacing a with a^1 and d with d^1 we find a new set of peaks, but in the same horizontal position. Now the second Ψ-line of the new set of peaks joins up with the first Ψ-line of the old peaks and it is clear how to walk down diagonally until the bottom line is reached.

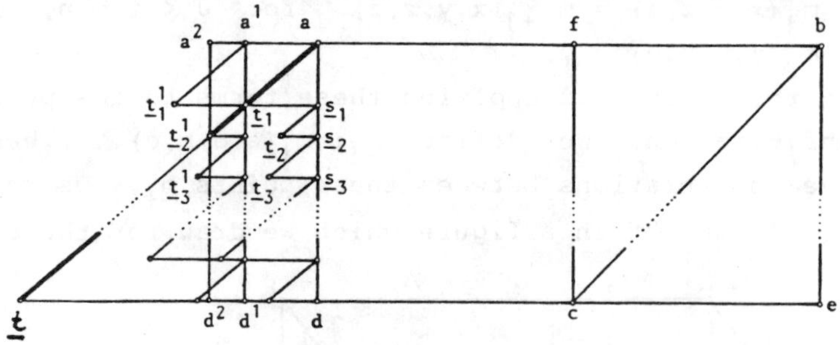

Since all points are constructed by applying Day's terms to the original points it is clear that syntactically the above method results in a continued substitution of terms into other terms, yielding a final term t(x,y,z) with the required property.

Coordinatization

We shall start with the simple idea of coordinatizing an affine line l using an abelian group. After embedding l into a plane and using two more mutually non-parallel lines l' and l", the addition x+y may easily be performed as shown in the picture below. Note that beforehand we must have selected an arbitrary element 0 on l to serve as the zero for addition.

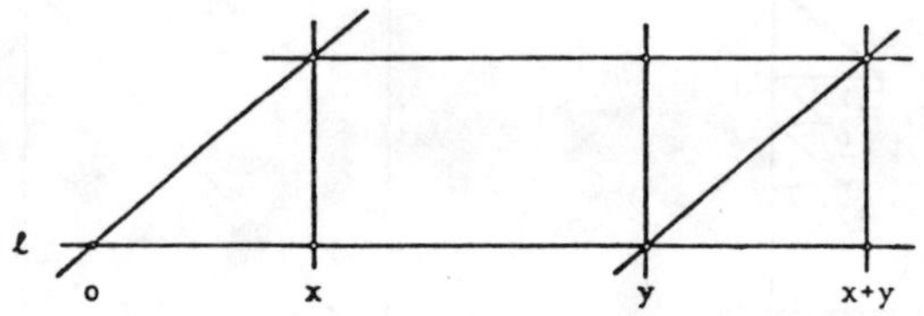

We now assume that an algebra A in a modular variety is given. In our analogy then the A will be the line l which we embed into the "plane" $A \times A$. Note that in $A \times A$ we may represent A as $\{(0,x) \mid x \in A\}$ or as $\{(x,0) \mid x \in A\}$, both being lines corresponding to the projection congruences π_1, resp. π_2. 0 is chosen arbitrarily in A. For the third line l", the choice is not quite as obvious. Since the geometrical operations we want to perform are parallel shifts and finding points of intersection, we need a congruence relation Δ such that every Δ-line intersects every π_1-line and every π_2-line. An obvious candidate for l" seems to be the diagonal $d = \{(x,x) \mid x \in A\}$ and therefore Δ should be at least the congruence relation generated by d in $A \times A$,

$$\Delta := < \{((x,x),(y,y)) \mid (x,y) \in A \times A\} >_{A \times A} .$$

Now, however, d need not be a line by itself, but the class of Δ containing d might be some $\bar{d}$ properly containing d. Such a $\bar{d}$ will certainly intersect each π_1- and each π_2-line, and if A is contained in a modular variety, it follows from the modified parallelogram principle that so will each other line of Δ. Finally, to construct x+y as above, points of intersections ought to be unique. This is the only hypothesis which is not guaranteed by modularity, but we shall see soon how to handle this case. Let us now for a moment assume that points of intersection are unique. It is easy to see with the shifting lemma that this is equivalent to requiring that d itself is a line, i.e. d = d. Now the familiar definition of x+y may be given and its properties developed. Proving associativity, commutativity and cancellativity leads to versions of exactly the geometrical configuration theorems which we met earlier. Since addition is defined via congruence relations and each such congruence relation is preserved by all polynomials of A it is not surprising that every polynomial of A is a homomorphism with respect to the ternary operation x-y+z. (Considering x-y+z instead of the group operations x+y and -x is preferable in any case, since it removes the discussion about choosing some arbitrary element for a zero.) <u>Thus A is polynomially equivalent to a module over some ring R.</u> Moreover all the algebras A in a fixed modular variety where d is a congruence class do form a subvariety of modules, the variety of "abelian algebras".

In the case that d by itself is not a line, intuitively $\bar{d}$ becomes too "thick" so that $\bar{d}$ intersects horizontal and vertical lines in more than one point, and as a consequence x+y cannot be defined uniquely. The obvious idea is to factor A by this "thickness" of $\bar{d}$. The idea indeed works thanks to the shifting lemma. The definition

$$b\,[1,1]\,c \iff \exists x\ (x,b)\ \Delta\ (x,c)$$

yields a congruence relation $[1,1]$ on A so that the coordinatization of $A/[1,1]$ works as described above, and $A/[1,1]$ is an abelian algebra.

More generally, given two congruences Θ and Ψ, we may reuse our trick of replacing $A\times A$ by its subalgebra Θ and we may replace d by those pieces of the diagonal which are congruent modulo Ψ to obtain:

$$\Delta_\Theta^\Psi := \langle\{((x,x),(y,y)) \mid (x,y)\in\Psi\}\rangle_\Theta$$

This is a congruence relation on the algebra Θ. Then, just as before we set

$$b\,[\Theta,\Psi]\,c \iff \exists x\ (x,b)\ \Delta_\Theta^\Psi\ (x,c)$$

and get that $[\Theta,\Psi]$ becomes a congruence relation on A, which is called the <u>commutator of Θ and Ψ</u>. This congruence multiplication was invented by J.D.H.Smith /20/, J.Hagemann and C.Herrmann /15/ and further completed and reinterpreted in /12/ and /7/. It reduces precisely to the familiar notion of commutators in groups and to the notion of ideal multiplication in rings. In distributive varieties the commutator of Θ and Ψ equals their intersection. All the notions and analogies from those theories, like primeness, nilpotency and solvability become available in general for modular varieties. They have provided a wealth of new insight into their structure theory. As an example we mention the following theorem which generalizes the famous Jonsson-lemma /18/. The essential idea for such a generalization is due to Hrushovskii, the present form is from /10/. We define $\sqrt{A}$ as the intersection of all prime congruences of A, and assume that A is in the (modular) variety generated by some class <u>K</u>. Then

<u>$A/\sqrt{A}$ is in $P_S HSP_U(K)$</u>.

In a distributive variety, primeness reduces to finitely subdirect irreducibility, so here $\sqrt{A} = 0$. Clearly, if A is abelian then no congruence relation will be prime, since $[\Theta,\Psi] = 0$ for all congruences Θ and Ψ. Thus $\sqrt{A} = 1$. In this way we have identified two "ends" of any modular variety and indeed, if certain restrictions are placed on the variety, it frequently will split into those two ends. Numerous results have been obtained in this direction, we only mention /11/ and the work of R.Freese and R.McKenzie /7/ and S.Burris and R.McKenzie /3/, but a general decomposition theory still has to be found.

References

1 H.Bauer, R.Wille: A proof of Hashimoto's refinement theorem. THD preprint 734, Darmstadt 1983.

2 S.Bulman-Fleming, A.Day, W.Taylor: Regularity and modularity of congruences. Algebra Universalis 4 (1974), 58-69.

3 S.Burris, R.McKenzie: Decidable varieties with modular congruences lattices. Memoirs of the A.M.S., 32 (1981), Number 246.

4 A.Day: A characterization of modularity for congruence lattice of algebras. Canad. Math. Bull. 12 (1969), 167-173.

5 W.Feit: An interval in the subgroup lattice of a finite group which is isomorphic to M_7. Algebra Universalis, to appear.

6 R.Freese, B.Jonsson: Congruence modularity implies the Arguesian identity. Algebra Universalis 7 (1977), 191-194.

7 R.Freese, R.McKenzie: Residually small varieties with modular congruence lattices. Transactions of the A.M.S., 246 (1981), 419-430.

8 H.P.Gumm: Über die Lösungsmengen von Gleichungssystemen über allgemeinen Algebren. Math.Z. 162 (1978), 51-62.

9 H.P.Gumm: The Little Desarguesian Theorem for modular varieties. Proc. of the A.M.S. 80 (1980), 393-397.

10 H.P.Gumm: Geometrical methods in congruence modular algebras. Memoirs of the A.M.S. , 45 (1983), Number 286.

11 H.P.Gumm: Congruence modularity is permutability composed with distributivity. Arch.d.Math. 36 (1981), 569-576.

12 H.P.Gumm: An easy way to the commutator in modular varieties. Arch.d. Math. 34 (1980), 220-228.

13 H.P.Gumm, C.Herrmann: Algebras in modular varities: Baer refinements, cancellation and isotopy. Houston J. of Math. 5 (1979), 503-523.

14 J.Hagemann: On regular and weakly regular congruences. THD preprint, Darmstadt 1973.

15 J.Hagemann, C.Herrmann: A concrete ideal multiplication for algebraic systems and its relation to congruence distributivity. Arch.d.Math. 32 (1979), 234-245.

16 J.Hashimoto: On direct product decomposition of partially ordered sets. Ann. of Math. 54 (1951), 315-318.

17 B.Jonsson: On the representation of lattices. Math. Scand. 1 (1953), 193-206.

18 B.Jonsson: Algebras whose congruence lattices are distributeve. Math. Scand. 21 (1967), 110-121.

19 P.P.Palfy, P.Pudlak: Congruence lattices of finite algebras and intervals in subgroup lattices of finite groups. Algebra Universalis 11 (1980), 22-27.

20 J.D.H.Smith: Mal'cev varieties. Lecture Notes in Math., Springer Verlag, Berlin - New York, 1976.

21 R.Wille, Kongruenzklassengeometrien. Lecture Notes in Math., Vol.113, Springer Verlag, Berlin - New York, 1970.

Gesellschaft für Strahlen- und
Umweltforschung mbH München (GSF)
Instit. für Medizinische Informatik
und Systemforschung (MEDIS)
Ingolstädter Landstr. 1
D-8042 Neuherberg, West-Germany

Universal Algebra and its Links with Logic,
Algebra, Combinatorics, and Computer Science
Proc. "25. Arbeitstagung über Allgemeine Algebra", Darmstadt 1983
P. Burmeister et al. (eds.)

29 – 40

INTERPOLATION IN UNIVERSAL ALGEBRA

H. K. Kaiser

In this article we try to give an account of the origin, development and possible future research in the field of interpolation in universal algebra.

Boolean algebras are of interest to various different types of mathematicians. Logicians for example are interested in them because of truth functions, computer scientists consider them when dealing with switching circuits.

In 1953 A.L.Foster started to investigate a special property of Boolean algebras of order 2 from the point of view of universal algebra by introducing the following concept:

A universal algebra $\langle A,\Omega\rangle$ is called primal if every finitary function on A with values in A is a termfunction over A.

Let $\langle A,\Omega\rangle$ be a universal algebra and k a fixed positive integer. On the set $F_k(A)$ of all functions $f:A^k\to A$ we define the operations $\omega\in\Omega$ pointwise. In this way we obtain the full k-ary function algebra $\langle F_k(A),\Omega\rangle$ over A. The subalgebra $\langle T_k(A),\Omega\rangle$ of $\langle F_k(A),\Omega\rangle$, generated by the k projections is called the algebra of k-ary termfunctions. Thus an algebra is primal iff $F_k(A)=T_k(A)$ for all $k\in\mathbb{N}$. Primal algebras with an at most countable family of fundamental operations are necessarily finite. An important characterization of primal algebras is the following structure theorem:

Theorem: A universal algebra A is primal iff the following conditions hold:

(i) A is simple,

(ii) A has no proper subalgebras,

(iii) A has no proper automorphisms,

(iv) A generates a congruence permutable variety,

(v) A generates a congruence distributive variety.

Examples of primal algebras are all prime fields and the Boolean algebras of order 2 (the trivial case of one-element algebras is excluded from the discussion throughout the paper). But there are many more examples. Davies showed the following result: Let N(n) denote the number of groupoids of cardinality n and P(n) the number of primal groupoids of cardinality n, then $P(n)/N(n) \to \frac{1}{e}$ with $n \to \infty$.

The concept of primality has been generalized in various directions. One way of doing this is starting with weakening the conditions of the structure theorem. For example, A.F.Pixley introduced the concept of quasiprimal algebras:

A finite algebra A is called quasiprimal, if the following holds:

(i) every subalgebra of A is simple,

(ii) A generates a congruence permutable variety,

(iii) A generates a congruence distributive variety.

This concept can be characterized via subalgebra of the full function algebra:

Theorem: A finite universal algebra A is quasiprimal iff every finitary function $f: A^n \to A$ $(n \geq 1)$ which preserves all isomorphisms between subalgebras of A is a term function.

Examples of quasiprimal algebras are the finite fields. Murskii showed that almost all groupoids are quasiprimal.

Of the many generalizations of primality we only mention one more: (which can be motivated by the fact, that cyclic groups of prime order are not quasiprimal, but fit into the following concept:)

A finite algebra A is called paraprimal, of all subalgebras of A are simple and A generates a congruence permutable variety.

E.Kiss showed that paraprimality cannot be characterized by preservation properties of termfunctions.

Another direction of generalizing primality takes as a starting point the equation $F_k(A)=T_k(A)$. We alter the right hand side by replacing the algebra of k-ary termfunctions by the algebra $P_k(A)$ of k-ary polynomial functions. $\langle P_k(A),\Omega\rangle$ is the subalgebra of the full k-ary function algebra, generated by the k projections and the k-ary constant functions.

We say: A is k-functionally complete, if $P_k(A)=F_k(A)$. A is called functionally complete, if $P_k(A)=F_k(A)$ for all positive intergers k.

All finite fields are functionally complete, and so are e.g. all finite simple non-abelian groups.

Functionally complete algebras are simple and - if their system of fundamental operations is at most countable - finite.

To overcome the obstacle of simplicity we define: $\langle A,\Omega\rangle$ is called affine complete, if every finitary function which is compatibel with all congruence relations on A is a polynomial function. If $C_k(A)$ denotes all k-ary compatibel functions over A, then the affine completeness of A can be expressed by $C_k(A)=P_k(A)$ for all $k\in\mathbb{N}$ (in this case the left hand side of our equation has been altered). Nontrivial examples of affine complete algebras are vectorspaces of dimension two.

Since no nontrivial lattice enjoys the property of functional completeness (every polynomial function necessarily preserves the order relation) a weakened from of "completeness" is used:
A lattice V is orderpolynomially complete, if every orderpreserving function is a polynomial function.

All these concepts are special forms of a general concept of completeness ([19]):

Let $\mathbb{K}$ be a class of universal algebras of the same type. For all $A\in K$ let $U_k(A)$, $V_k(A)$ be subsets of $F_k(A)$ such that $U_k(A)\subseteq V_k(A)$. A is called k-(U,V)-complete if $U_k(A)=V_k(A)$, and A is called (U,V)-complete, if this equation holds for all $k\in\mathbb{N}$.

The most important special form of completeness for us is the so called interpolation property. In many practical problems one has to calculate with functions over some fixed algebra. This is made easier if one has some sort of algebraic characterization of the functions in question, e.g. a representation by a polynomial function. Often it is enough to know some representation on any finite number of points at the domain of the function.

So one looks for algebras where every finitary function has this property, in other words: every function can be "interpolated" by polynomial functions. So one introduces a local version of functional completeness:

Let A be a universal algebra, $f:A^k \to A$ and n some positive integer. If for every $N \subseteq A^k$, $|N| \leq n$, there is a $p \in P_k(A)$ such that $f/N=p/N$ then we say that f has the n-interpolation property. The set of all $f \in F_k(A)$ with the n-interpolation property is denoted by $L_nP_k(A)$, and we put $LP_k(A) := \bigcap_{n \in \mathbb{N}} L_nP_k(A)$.
A is said to have the n-interpolation property, if $F_k(A)=L_nP_k(A)$ for all $k \in \mathbb{N}$ and to have the interpolation property if these equations holds for all $n \in \mathbb{N}$.
Obviously, all fields have the interpolation property. We even have an algorithm to find for any finite cardinality of dom f the representing polynomial function (interpolation formulas of Lagrange and Newton).

Algebras with the interpolation property are necessarily simple. If all binary functions over A have the interpolation property ($LP_2(A)=F_2(A)$) then A has the interpolation property. This result is derived from the fact that all finitary functions over A can be obtained from binary functions over A via composition. If there is a surjective binary polynomial function over A, depending on both arguments, then $LP_1(A)=F_1(A)$ implies the interpolation property of A.

At first the interpolation property of given universal algebras has been checked by ad hoc methods. Then various characterizations of the interpolation property have been found. Here some examples:
Theorem: A universal algebra A has the interpolation property iff the ternary discriminator function $t:A^3 \to A$, defined by: $t(x,y,z)=z$ for $x=y$ and $t(x,y,z)=x$ otherwise, has the interpolation property.
A universal algebra A, $|A| \geq 3$, has the interpolation property iff the dual discriminator $d:A^3 \to A$, defined by: $d(x,y,z)=x$ for $x=y$ and $d(x,y,z)=z$ otherwise, has the interpolation property.
In general it is hard to find the interpolating polynomial functions for the discriminator and the dual discriminator. Hence - for the problem of finding all algebras with the interpolation property in a given variety - one tries to find characterizations which are easier to apply:

Theorem: A universal algebra A has the interpolation property iff the following conditions hold:

(i) A is simple,

(ii) there exists $a \in A$ and $p,t \in LP_2(A)$ such that:

(α) $p(x,x)=a$ for all $x \in A$,

(β) $t(p(x,y),y)=x$ for all $x,y \in A$,

(iii) there exists $q \in LP_2(A)$ such that:

(α) q is not constant,

(β) $q(x,a) = q(a,x)=a$ for all $x \in A$.

As an application we determine all groups $\langle G,+,-,0 \rangle$ with the interpolation property. For $p(x,y)$ we take $x-y$, for $t(x,y)$ we take $x+y$, for a the neutral element 0 and for non-abelian groups the commutator $[x,y]$ satisfies property (iii). It is easy to show that in case of commutative groups every binary function with the interpolation property which satisfies condition (iii)(β) is necessarily constant. Hence we have:

Theorem: In the variety of groups every simple group either has the interpolation property or is abelian.

Other examples of algebras with the interpolation property can be found in [17],[18].

A final example of a tool which can be used to check the interpolation property of an algebra is the "interpolation lemma" ([11]):

Let A be a universal algebra. Two k-tuples $(a_1,\ldots,a_k)$, $(b_1,\ldots,b_k) \in A^k$ are said to be of the same pattern if $a_i=a_j$ iff $b_i=b_j$. $f:A^k \to A$ is called a pattern function, if $f(a_1,\ldots,a_k) \in \{a_1,\ldots,a_k\}$ for all $(a_1,\ldots,a_k) \in A^k$ and if $f(a_1,\ldots,a_k)=a_i$ and $(b_1,\ldots,b_k)$ is of the some pattern as $(a_1,\ldots,a_k)$, then $f(b_1,\ldots,b_k)=b_i$. A semi-pattern function differs from a pattern function in that nothing is known of its values if all arguments are different. A semi-near-projection is a function f which reduces to projections when identifying any two of its variables. If we always obtain the same projection x_i by this then f is called a semi-projection (onto x_i). A function is said to be non-trivial, if it is not a projection.

Interpolation Lemma: Let A be an algebra, $|A|\geq 3$, and suppose that there exists a family of functions with the interpolation property $f_\lambda(x_1,x_2,\ldots)$ $(\lambda\in\Lambda)$ over A such that for all $\lambda\in\Lambda$:

(i) f_λ is a semi-near-projection,

(ii) $f_\lambda(x_1,x_1,x_3,\ldots)=x_1$,

(iii) f_λ is not a semi-projection onto x_2,

further

(iv) for all distinct $a,b,c\in A$ there exists an f_λ, say of arity n, and elements $d_3,\ldots,d_n\in A$ such that $f_\lambda(a,b,d_3,\ldots,d_n)=c$.

Then A has the interpolation property.

From this result one can easily derive e.g. the characterization of the interpolation property by the dual discriminator.

All these results have the disadvantage of being mere existence theorems. They give us no hint how to actually find to a given function the corresponding interpolating polynomial functions.

There has already been made some progress of finding interpolation algorithms for special algebras (apart from the case of fields and the Boolean algebra of order 2). There have been found interpolation algorithms in the case of special simple non-commutative rings ([22]), of special classes of simple near-rings ([43]) and of some simple nonabelian groups ([33],[34]).

But algebras with the interpolation property have also been successfully used in studying and classifying simple algebras in given varieties. The first theorem of this kind was shown by Gumm and McKenzie ([15],[39]) and is the extension of the previous theorem characterizing groups with the interpolation property to congruence permutable varieties:

A function $f\in F_k(A)$ is called affine with respect to an abelian group $\langle A,+,-,0\rangle$ if for all $x_1,\ldots,x_k, y_1,\ldots,y_k\in A$ the following equation holds:

$f(x_1,\ldots,x_n)+f(y_1,\ldots,y_k)=f(x_1+y_1,\ldots,x_k+y_k)+f(0,\ldots,0)$.

$\langle A,\Omega\rangle$ is called affine with respect to an abelian group $\langle A,+,-,0\rangle$ if every $\omega\in\Omega$ is affine.

Theorem: Let $A=\langle A,\Omega\rangle$ be a simple algebra in a congruence permutable variety. Then A either has the interpolation property or is affine with respect to an abelian group $\langle A,+,-,0\rangle$ which is either an elementary p-group, p-prime, or torsion free.

Hence one immediately knows that a simple finite algebra in a congruence permutable variety has the interpolation property if its cardinality is not a power of a prime number.

Other theorems of the Gumm-McKenzie type can be found in [4], [27], [51],[52],[54] and [55].

Another development closely connected to interpolation has been started just recently, namely the discussion of ideas of apparoximation theory on the level of universal algebra (see [31],[32]):

We start with a topological algebra $\langle A,\Omega,\tau\rangle$(to avoid trivialities we assume that τ satisfies the separating axiom T_2). We endow $F_k(A)=A^{(A^k)}$ with the product topology and say:

A topological universal algebra A has the approximation property if $P_k(A)$ is dense in $F_k(A)$ for all $k\in\mathbb{N}$.

A typical result obtained is the following:

Let A be a topological group with T_2. Then A has the approximation property iff A is nonabelian such that all nontrivial normal subgroups are dense in A.

Future research in the field of interpolation will possibly give attention to interpolating algorithms for special algebras, and develop an approximation theory in topological universal algebras.

One goal will be to introduce methods of interpolation to partial algebras and heterogenous algebras, to make interpolation a working tool for computer scientists.

Author's adress:
Institut für Algebra
Technische Universität Wien
Argentinierstraße 8
A-1040 WIEN, AUSTRIA

REFERENCES

[1] K.BAKER - A.F.PIXLEY: Polynomial interpolation and the Chinese remainder theorem for algebraic systems, Math.Z. 143, 165-174, 1975.

[2] D.CLARK - P.KRAUSS: Paraprimal algebras,Alg.Univ.6, 165-192, 1976.

[3] D.CLARK - P.KRAUSS: Varieties generated by para primal algebras, Alg.Univ.7, 93-114, 1977.

[4] B.CSAKANY: Homogeneous algebras are functionally complete. Preprint.

[5] R.O.DAVIES: On n-valued Sheffer functions, Preprint, Univ.of Leicester, 1975.

[6] D.DORNINGER - G.EIGENTHALER: On compatible and order-preserving functions on lattices. Preprint.

[7] D.DORNINGER - W.NÖBAUER: Local polynomial functions on lattices and universal algebras. Coll.Math., XLII, 83-93, 1979.

[8] A.L.FOSTER: Generalized "Boolean" theory of universal algebra. Math.Z.59, 191-199, 1953.

[9] A.L.FOSTER: Functional completeness in the small. Math.Ann.143, 29-58, 1961.

[10] E.FRIED: Interpolation in weakly associative lattices. Studia Sci.Math. 13, 193-200, 1978.

[11] E.FRIED - H.K.KAISER - L.MARKI: An elementary approach to polynomial interpolation in universal algebras. Alg.Univ. 15, 40-57, 1982.

[12] E.FRIED - A.F.PIXLEY: The dual discriminator function in universal algebra. Acta Sci.Math.41, 83-100, 1979.

[13] G.GRÄTZER: On Boolean functions (Notes on lattice theory II), Revue de Math. Pure et Appl.7, 693-697, 1962.

[14] P.A.GROSSMANN - H.LAUSCH: Interpolation in semilattices. Preprint.

[15] H.P.GUMM: Algebras in congruence permutable varieties: geometrical properties of affine algebras. Alg.Univ.9, 8-34, 1979.

[16] H.HULE - W.NÖBAUER: Local polynomial functions on universal algebras.
An.Acad.Brasil de Cien. 49, 365-372, 1977.

[17] M.ISTINGER - H.K.KAISER: A characterization of polynomially complete algebras.
J.of Algebra 56, 103-110, 1979.

[18] M.ISTINGER - H.K.KAISER - A.F.PIXLEY: Interpolation in congruence permutable algebras.
Coll.Math.XLII, 229-239, 1979.

[19] H.K.KAISER: Der Defekt endlicher abelscher Gruppen.
Monatsh.Math.78, 109-116, 1974.

[20] H.K.KAISER: A class of locally complete universal algebras.
J.London Math.Soc.9, 5-8, 1974.

[21] H.K.KAISER: On a problem in the theory of primal algebras.
Algebra Univ.5, 307-311, 1975.

[22] H.K.KAISER: Über das Interpolationsproblem in nichtkommutativen Ringen.
Acta Sci.Math.36, 271-273, 1974.

[23] H.K.KAISER: Über lokalpolynomvollständige universale Algebren.
Abh.Math.Sem.Univ.Hamburg, 43, 158-165, 1975.

[24] H.K.KAISER: A note on simple universal algebras.
Coll.Math.Esztergom, 437-440, 1977.

[25] H.K.KAISER: A remark on algebras with the interpolation property.
Contributions to general algebra I, 113-1119, 1978.

[26] H.K.KAISER - R.LIDL: Erweiterungs- und Redeipolynomvollständigkeit universaler Algebren.
Acta Math.Acad.Sci.Hung.26, 251-257, 1975.

[27] H.K.KAISER - L.MARKI: Remarks on a paper of L.Szabo and A.Szendrei.
Acta Sci.Math.42, 95-98, 1980.

[28] E.KISS: Termfunctions and subalgebras.
Preprint.

[29] R.A.KNOEBEL: Primal extensions of universal algebra.
Math.Z.115, 53-57, 1970.

[30] J.KOLLAR: Interpolation in semigroups.
Preprint

[31] G.KOWOL: Approximation durch Polynomfunktionen auf universellen Algebren.
Monatsh.Math.93,1982.

[32] G.KOWOL: Approximation on universal algebras by means of polynomial functions.
Preprint

[33] H.LAUSCH: Interpolation on alternating groups of prime degree.
Algebra Paper, Nr.34, Monash Univ.,1978.

[34] H.LAUSCH: Interpolation on alternating groups and on PSL (2,p).
Algebra Paper Nr. 35, Monash Univ., 1978.

[35] H.LAUSCH - W.NÖBAUER: Funktionen auf endlichen Gruppen.
Public.Math.23, 53-61, 1976.

[36] L.MARKI: Über einige Vollständigkeitsbegriffe in Halbgruppen und Gruppen.
Acta Math. Acad.Sci.Hun. 26, 343-346, 1975.

[37] W.D.MAURER - J.L.RHODES: A property of finite simple nonabelian groups.
Proc.Amer.Math.Soc.16, 552-554, 1965.

[38] R.McKENZIE: On spectra, and the negative solution of the decision problem for identities having a finite nontrivial model.
J.Symb.Logic.40, 186-195, 1975.

[39] R.McKENZIE: On minimal, locally finite varieties with permutable congruence relations.
Preprint, Berkeley, 1976.

[40] R.McKENZIE: Para primal varieties: A study of finite axiomatizability and definable principal congruences in locally finite varieties.
Alg.Univ.8, 336-348, 1978.

[41] V.L.MURSKII: The existence of a finite basis of identities, and other properties of "almost all'" finite algebras.
Problemy Kibernet.30, 43-56 (Russian), 1975.

[42] W.NÖBAUER: Compatible and conservative functions on residue-class rings of the integers.
Coll.Math.Debrecen, 245-257, 1974.

[43] R.PFISTER: Interpolation in Fastringen.
Diss.TU Wien, Juni 1981

[44] G.PILZ: Near-rings of compatible functions.
Preprint.

[45] A.F.PIXLEY: Functionally complete algebras generating permutable and distributive classes.
Math.Z.114, 361-372, 1970.

[46] A.F.PIXLEY: The ternary discriminator function in universal algebra.
Math.Ann.191, 167-180, 1971.

[47] A.F.PIXLEY: Completeness in arithmetical algebras.
Alg.Univ.2, 179-196, 1972.

[48] A.F.PIXLEY: A survey of interpolation in universal algebra.
Proc.Coll.Esztergom, 583-607, 1977.

[49] R.W.QUACKENBUSH: Structure theory for equational classes generated by quasi-primal algebras.
Transactions AMS 187, 127-145, 1974.

[50] I.G.ROSENBERG: Über die funktionale Vollständigkeit in den mehrwertigen Logiken.
Rozpravy Cesk.Akad.Ved.Ser.Math.Nat.Sci.,80,3-93,1970.

[51] I.G.ROSENBERG - D.SCHWEIGERT: Local Clones.
Preprint.

[52] I.G.ROSENBERG - L.SZABO: Local Completeness I.
Preprint.

[53] D.SCHWEIGERT: Über endliche ordnungspolynomvollständige Verbände.
Monatsh.Math.78, 68-76, 1974.

[54] L.SZABO: Tolerance-free algebras with a majority function.
Preprint.

[55] L.SZABO - A.SZENDREI: Almost all algebras with triply transitive automorphism groups are functionally complete.
Acta Sci.Math., 41, 391-402, 1979.

[56] W.TAYLOR: Characterizinh Mal'cev conditions.
Alg.Univ.3, 351-397, 1973.

[57] H.WERNER: Produkte von Kongruenzklassengeomtrien.
Math.Z.121, 111-140, 1971.

[58] H.WERNER: Congruences on products of algebras and functionally complete algebras.
Alg.Univ.4, 99-105,1974.

[59] H.WERNER: Discriminator algebras.
Studien zur Algebra, Bd.6, Akademie-Verlag, Berlin, 1978.

[60] J.WIESENBAUER: Interpolation on modules.
Contr.to general algebra I, 399-404, 1978.

[61] R.WILLE: Kongruenzklassengeometrien.
Lecture Notes in Math.113, Springer, Berlin-Heidelberg-New York, 1970.

[62] R.WILLE: Eine Charakterisierung endlicher, ordnungspolynomvollständiger Verbände.
Arch.Math.(Basel)28, 557-560, 1977.

[63] R.WILLE: A note on algebraic operations and algebraic functions of finite lattices.
Houston J.Math.3, 593-597, 1977.

Universal Algebra and its Links with Logic,
Algebra, Combinatorics, and Computer Science
Proc. "25. Arbeitstagung über Allgemeine Algebra", Darmstadt 1983
P. Burmeister et al. (eds.)

41 – 66

ALGEBRAIC AND DEDUCTIVE CONSEQUENCE OPERATIONS

W. Felscher, J. Schulte Mönting

Many logical calculi may be considered as being defined by a certain class of abstract algebras which, at the same time, are partially ordered sets with a largest element as the designated truth value. A conceptual frame in which to discuss such situations, was developed by Rasiowa-Sikorski [53] and later refined by Bammert [68] and Rasiowa [74] . More recently, much effort was spent on further generalizations, permitting to include new examples into the abstract approach; we mention here only Czelakowski [81] and Blok-Köhler-Pigozzi [82] .

The present article had its origin in observations which occurred to us while repeatedly teaching courses on classical and non-classical logic. It was our idea to obtain further insights, into the general machinery as well as into particular examples, by making systematical use of notions from universal algebra. As we wanted to stay reasonably close to the traditional examples, we decided to avoid the additional apparatus necessary for perfect generality; thus we assumed throughout that our algebras carry, aside from possible further operations, a binary operation representing implication and satisfying the laws defining implicative algebras in the sense of Rasiowa [74] . As for our results, most of them answer to questions arising in the context of universal algebra. However, in the second part of § 4 we shall present certain purely syntactical methods for the construction of deductive calculi.

In order to make the article easily accessible, we have included the introductory § 1 . We hope that it may serve as an introduction to certain ideas from logic to readers familiar with universal algebra, while logicians may appreciate the occasional inspiration offered by an algebraic point of view.

In the universal-algebraic approach, every suitable class A of algebras determines an *algebraic* consequence operation cn_{A} . Our main tool in studying the relationship between A and cn_{A} is the class of *test algebras* of cn_{A} (the S-algebras of Rasiowa-Sikorski), and we show in Theorem 2 that this class can be described as the class $\mathsf{T}\mathsf{A}$ generated by A under certain algebraic constructions; these constructions are closely connected with a *representation property*, as it is familiar from the case of classical logic and Boolean algebras. The first object of our investigation then are the *equational* consequence operations cn_{A} , i.e. those which can be computed by an algebraic calculus of equations based on the equations determined by A (such as the consequences in classical logic can be computed by the calculus of Boolean equations). In Theorem 1 and in the corollary to Theorem 2 we obtain simple criteria characterizing equational consequence operations; it may be mentioned that early results in this area were presented by the first author at the Arbeitstagung in Darmstadt in June 1973 . The second object of our investigation are the consequence operations cn_{A} which can be defined by *axioms* and *finitary rules*, and for classes A such that $\mathsf{A} = \mathsf{T}\mathsf{A}$ this syntactical definability of cn_{A} is equivalent to the algebraic property that A is quasiequational; the various connections between syntactical and algebraical properties are collected in the Theorems 3 - 6 . — The third object to be investigated are the *deductive* consequence operations cn_{A} , i.e. those which admit the so-called deduction rule; for classes A such that $\mathsf{A} = \mathsf{T}\mathsf{A}$ we obtain algebraic characterizations of deductivity in Theorem 8 . Analyzing the proof of one of these characterizations, we arrive at a purely syntactical characterization of the *smallest* deductive consequence operation containing a given (arbitrary) cn_{A} . These results on deductive consequence operations were presented by the second author at the conference on Mathematical Logic at Oberwolfach in April 1980 ; recently a related, considerably more general characterization of deductivity has been obtained by Blok-Köhler-Pigozzi [82] .

Having proceeded so far by an algebraic approach, we continue with purely syntactical arguments. Making use of the process of *relativization* of defining axioms and rules, we prove a much simpler characterization of the smallest deductive consequence containing a given one than was afforded by the algebraic approach. Relativization here consists in taking a new variable u_o and replacing every formula v , occurring as an axiom, as a premiss or as a conclusion of a rule, by the formula $u_o \to v$; in this manner, various examples of seemingly accidental formulations of deductive calculi obtain a well-defined meaning.

§ 1 BASIC CONCEPTS

ALGEBRAS. We consider (abstract) algebras with finitary operations, and we distinguish between an algebra A and its underlying set u(A) ; an algebra is *singular* if it contains only one element. We assume that every algebra A determines uniquely the sequence $<f_i^A \mid i\varepsilon I>$ of its operations and, thereby, also the sequence $\Delta = <n_i \mid i\varepsilon I>$ of natural numbers n_i such that n_i is the arity of the operation f_i^A ; Δ is called the *signature* of A. Consider now algebras of a fixed signature. For every class A of algebras define A_e to be the class of all algebras A such that A is in A or A is singular. For every class A we define the classes $\mathsf{I}\mathsf{A}$, $\mathsf{H}\mathsf{A}$, $\mathsf{S}\mathsf{A}$, $\mathsf{P}\mathsf{A}$ of all isomorphic images, homomorphic images, subalgebras and direct products of algebras in A respectively; we also define classes $\mathsf{L}_0\mathsf{A}$, $\mathsf{L}_1\mathsf{A}$ by

$B \varepsilon \mathsf{L}_0\mathsf{A}$ if every subalgebra of B, generated by a finite set, belongs to A,

$B \varepsilon \mathsf{L}_1\mathsf{A}$ if every subalgebra of B, generated by an at most countably infinite set, belongs to A.

An algebra B is A-*representable* if, for any two distinct elements b,b' of B, there exists an A in A and a homomorphism f from B into A such that $f(b) \neq f(b')$. If R_f is the congruence relation on B, induced by f, then the quotient B/R_f is in ISA, and if H is the set of all these R_f then B can be isomorphically embedded into $\Pi <B/R_f \mid R_f \varepsilon H>$. Thus the class RA of all A-representable algebras is ISPISA_e.

EQUATIONS AND QUASIEQUATIONS. Let T be a term algebra of signature Δ, generated absolutely freely by a set X of *variables*. A T-*equation* p is an ordered pair $[p_0,p_1]$ of terms in T ; a T-*quasiequation* [M,p] consists of a finite set M of T-equations and a single T-equation p. A quasiequation [M,p] *holds* in an algebra A if every homomorphism from T into A, which identifies all equations in M, also identifies p. A class A is *equational*, resp. *quasiequational*, if there exists a set of equations, resp. of quasiequations, such that A consists precisely of those algebras in which these equations, or quasiequations, hold. This definition depends only seemingly on the (cardinality of the) set X of variables, for if it is satisfied for *one* set X then it is satisfied also for *every infinite* set X. For every class A there exists a smallest equational class EA containing A, the *equational closure*, and there exists a smallest quasiequational class $\mathsf{Q}A$ containing A , the *quasiequational closure*; the class $\mathsf{E}A$ coincides with $\mathsf{HSPIS}A$. We later shall have occasion to study the classes $\mathsf{T}A$ and T_0A , defined as $\mathsf{T}_0A = \mathsf{L}_0\mathsf{R}A$, $\mathsf{T}A = \mathsf{L}_1\mathsf{R}A$, whence $\mathsf{T}A \subseteq \mathsf{T}_0A \subseteq \mathsf{Q}A \subseteq \mathsf{E}A$.

THE CLASS J . From now on, we consider a fixed term algebra T with a countably infinite set X of variables, and we assume that in Δ there occurs a particular arity $n_0 = 2$; we denote the corresponding binary operation in T by $\rightarrow$ and in any other algebra by $\supset$. Let J be the class of all algebras A for which u(A) carries a partial order $\leq$ with a largest element e_A and such that

$$a \leq b \quad \text{if and only if} \quad a \supset b = e_A \quad .$$

Then J is quasiequational, defined by the conditions

$$\begin{aligned} & a \supset a = b \supset b \quad (= e_A) \quad , \\ & a \supset e_A = e_A \quad , \\ & \text{if} \quad a \supset b = b \supset c = e_A \quad \text{then} \quad a \supset c = e_A \quad , \\ & \text{if} \quad a \supset b = b \supset a = e_A \quad \text{then} \quad a = b \quad . \end{aligned}$$

If Δ consists of n_0 only then J is the class of *implicative algebras* of Rasiowa [74] .

Throughout this article, all classes of algebras shall be subclasses of J , *and all algebras considered shall belong to* J *unless espressedly stated otherwise.*

FILTERS . Consider an algebra B which may not belong to J . If h is an homomorphism from B into an algebra A in J then the congruence relation R_h on B induced by h is completely determined by the congruence class $h^{-1}(e_A)$: it is $h(b) = h(b')$ equivalent to $h(b \supset b') = h(b' \supset b) = e_A$. For every class A the sets $h^{-1}(e_A)$, determined by homomorphisms from B into algebras A in A , are called the A-*filters* of B . In the particular case that B is the term algebra T , let ∇ be a term $x \rightarrow x$ for a fixed variable x ; then the A-filters of T are the sets $h^{-1}h(\nabla)$ for homomorphisms h from T into algebras in A .

AXIOMS AND RULES . Let E be a set. Let A be a subset of E , and let R be a set of functions r such that both def(r) and im(r) are sets of subsets of E . Let H be the family of all subsets Y of E which contain A and are closed with respect to all r in R : if $r \in R$ and $Z \subseteq Y$, $Z \in def(r)$ then $r(Z) \subseteq Y$. Then H is a closure system on E , and the corresponding closure operation is said to be *defined by* the set A of *axioms* and the set R of *rules*. — A function r is *finitary* if $G \in def(r)$, $v \in r(G)$ implies the existence of a finite subset F of G such that $v \in r(F)$; r is *monotonic* if, for G,G' in def(r), it follows from $G \subseteq G'$ that $r(G) \subseteq r(G')$. If all rules in R are finitary and monotonic then the closure operation defined by A and R is finitary. — For rules r , and also for closure operation c , statements of the form $v \in r(G)$ will often be written as $G \vdash v$ if it is clear which rule or which closure operation is meant; also, notations such as $g_0, \ldots, g_{n-1} \vdash v$ will be employed in the usual manner.

CALCULI. Let E be the set underlying the term algebra T . A closure operation c on E is said to be *structural* (Łos-Suszko [58]) if it commutes with all endomorphisms g of T , i.e. if $g(c(G)) \subseteq c(g(G))$ for every subset G of E ; a structural consequence operation shall be called a *calculus*.

A closure operation c , defined by axioms A and rules R , is a calculus if each of the rules in R is monotonic and commutes with all endomorphisms of T and if A is closed with respect to all these endomorphisms. For let g and G be given; the inclusion $g(c(G)) \subseteq c(g(G))$ will follow if it can be shown that the set D , consisting of all d such that $g(d) \in c(g(G))$, contains G and is closed with respect to c . Now $G \subseteq D$ is clear, and $A \subseteq D$ follows from $g(A) \subseteq A \subseteq c(0) \subseteq c(g(G))$; it thus remains to be shown that D is closed with respect to every rule r . Since r commutes with g and is monotonic, it follows from $g(D) \subseteq c(g(G))$ that $g(r(D)) \subseteq r(g(D)) \subseteq r(c(g(D)))$, and this implies $g(r(D)) \subseteq c(g(D))$ since c-closed sets certainly are r-closed. — This result can be stated also as follows: if c is the calculus defined by axioms A and rules R , and if each rule is monotonic and commutes with all endomorphisms, then the closure c(G) of a set G can be obtained by *first* closing up A under all endomorphisms to a set A^* , and *then* applying the rules from R to $A^* \cup G$: an additional *substitution rule*, producing closure under endomorphisms, needs to be applied to axioms *only*. For particular rules r , this observation is the content of Sobociński [74] and Lambros [79] .

MP - CALCULI. A closure operation c on the set E = u(T) is an *MP-map* if

$$v, v \to w \vdash w$$

holds for all terms v,w . This particular condition defines a rule which maps every set $\{v, v\to w\}$ into $\{w\}$. An MP-map is *pure* if it is the closure operation defined by this *one* rule and *some* set of axioms. Every pure MP-map is finitary. Every pure MP-map, defined by a set of axioms closed with respect to all endomorphisms, is a calculus. MP-maps which are calculi are called MP-*calculi*.

ALGEBRAIC CONSEQUENCE OPERATIONS . Let $\mathbf{A}$ be a class of algebras; let H be the closure system on the set E of terms, consisting of all $\mathbf{A}$-filters of T and all intersections of families of such filters. The corresponding closure operation $cn_{\mathbf{A}}$ is called the *algebraic consequence operation defined by* $\mathbf{A}$. For every subset G of E let $H(G,\mathbf{A})$ be the class of all homomorphisms h from T into algebras in $\mathbf{A}$ such that $G \subseteq h^{-1}h(\nabla)$; then

$$cn_{\mathbf{A}}(G) = \bigcap < h^{-1}h(\nabla) \mid h \in H(G,\mathbf{A}) > .$$

Making use of the correspondence between filters and congruence relations, we also obtain $cn_{\mathbf{A}}(G)$ as the congruence class of ∇ for the congruence relation

$$Q(G,\mathbf{A}) = \bigcap < R_h \mid h \in H(G,\mathbf{A}) > ,$$

where R_h is the congruence relation on T induced by h . — Every algebraic consequence operation is an MP-calculus.

We shall, in later paragraphs, prove various statements saying that certain consequence operations $cn_{\mathbf{A}}$ are finitary; in particular, $cn_{\mathbf{J}}$ *is* finitary. The following example shows that, in general, $cn_{\mathbf{A}}$ is *not* finitary. Consider the case that Δ contains only the operation $\to$. For every natural number n define the algebra A_n by $u(A_n) = \{i \mid o \leqq i \leqq n\}$ and $i \supset j = n$ if $i \leqq j$, and $i \supset j = o$ otherwise. Let $\mathbf{A}$ be the class of all A_n , $n \in \omega$. Assume that the variables of T have been enumerated as x_i , $i \in \omega$, and define terms $v_{ij} = x_i \to x_j$ for $i < j$ and $w_i = (x_{i+1} \to x_i) \to x_o$; let G be the set of all v_{ij} and w_i . *Then* x_o *is in* $cn_{\mathbf{A}}(G)$ *but is not in any* $cn_{\mathbf{A}}(F)$ *for finite subsets* F *of* G . For let h be a homomorphism from T into some A_n which satisfies all the v_{ij} ; let ν be the sequence from ω into A_n , defined by $\nu(i) = h(x_i)$. It follows from the construction of the v_{ij} that ν is monotonic; since n is finite, ν must eventually become constant with the value n . If m is such that $\nu(m) = \nu(m+1) = n$ and if h also satisfies w_m , then $n = h(w_m) = (n \supset n) \supset \nu(o) = n \supset \nu(o)$ implies $\nu(o) = n$; hence h satisfies x_o . On the other hand, let F be a finite subset of G and let k be an upper bound for the numbers i,j such that F consists of v_{ij} and w_i ; let h be a homomorphism from T into A_{k+1} such that $h(x_i) = i$ for $i \leq k+1$. Then h satisfies all the v_{ij} and w_i in F but, obviously, not x_0 . — This example can be generalized by substituting for ω a regular ordinal number Ω and for the natural numbers i,j,n ordinal numbers smaller than Ω . In that case, if T is a term algebra with Ω variables, then x_0 is in $cn_A(G)$ but is not in any $cn_A(F)$ for subsets F of G of smaller cardinality.

TEST ALGEBRAS . For closure operators c,d on E , define $c \subseteq d$ to mean that $c(G) \subseteq d(G)$ for every subset G of E . Consider now (1) a closure operation cs on E and (2) a consequence operation cn_A defined by a class **A** . We say that cs is *correct* (or *sound*) for cn_A if $cs \subseteq cn_A$; we say that cs is *complete* for cn_A if $cs = cn_A$. In order that cs be correct for cn_A it is necessary and sufficient that every **A**-filter on T be cs-closed. Consequently, *every* cs determines a *largest* subclass **S** of **J** such that cs is correct for cn_A : it is **S** the class of all A in **J** such that, for every homomorphism h from T into A , the set $h^{-1}h(\nabla)$ is cs-closed. The algebras in **S** were introduced in Rasiowa-Sikorski [53] under the name of S-*algebras* of cs ; we shall call them here the *test algebras* of cs . It follows from the maximality of **S** that this class is closed with respect to $\mathbf{I}, \mathbf{S}, \mathbf{P}, \mathbf{L}_1, \mathbf{R}, \mathbf{T}$. If cs is correct for *some* cn_A then it follows from $\mathbf{A} \subseteq \mathbf{S}$ that $cn_S \subseteq cn_A$; hence cn_S is the best approximation from above to cs , and if cs is complete for *any* cn_A at all then it certainly is complete for cn_S .

COMPLETENESS . Let cs be a closure operation on E and let G be a subset of E ; define S(G) to be the set of all [v,w] such that both $v \to w$ and $w \to v$ are in cs(G) . The following criterion is the content of Bammert [68] , Satz 4 :

A calculus cs is complete for the consequence cn_S defined by its test algebras if and only if

(i) every S(G) is a congruence relation on T ,

(ii) every quotient T/S(G) is in **J** ,

(iii) every set cs(G) is the congruence class of ∇ under S(G) .

These conditions are necessary since $cs = cn_S$ implies $S(G) = Q(G, \mathbf{S})$. Conversely, if $B = T/S(G)$ is in **J** then it will be sufficient to show that B is in **S** , for in that case the natural homomorphism π from T onto B is in $H(G, \mathbf{S})$ whence $cn_S(G) \subseteq \pi^{-1}\pi(\nabla) = S(G)(\nabla) = cs(G)$ by (iii) . It thus remains to be shown that, for every homomorphism h from T into B , the set $h^{-1}h(\nabla)$ is cs-closed.

To this end, it will suffice to find, for every v in $cs(h^{-1}h(\nabla))$, a subset D of $h^{-1}h(\nabla)$ and an endomorphism g of T such that v is in $cs(D)$ and the homomorphisms h and πg coincide for v and for the elements of D. For in that case it follows from $v \in cs(D)$ that $g(v) \in g(cs(D)) \subseteq cs(g(D))$, and $\pi g(D) = h(D) = h(\nabla) = \pi(\nabla)$ implies $g(D) \subseteq \pi^{-1}\pi(\nabla) = cs(G)$, hence $g(v) \in cs(g(D)) \subseteq cs(G)$, $h(v) = \pi g(v) = \pi(\nabla) = h(\nabla)$ and $v \in h^{-1}h(\nabla)$. — In order to find D and g, assume first that cs is finitary, and let D be a finite subset of $h^{-1}h(\nabla)$ such that $v \in cs(D)$; let Y be a finite subset of the set X of variables of T such that v and D belong to the subalgebra generated by Y in T. Then g can be defined on Y be choosing $g(y)$ to be an element from the congruence class $h(y)$. If cs is not finitary then we employ the axiom of choice in order to define g on X.

§ 2 EQUATIONAL CONSEQUENCE OPERATIONS

Consider a consequence operation cn_{A} defined by a class A. There is, in general, no algorithm permitting us to compute the consequences in $cn_{\mathsf{A}}(G)$ from the terms in G. Now we know that $cn_{\mathsf{A}}(G)$ is the congruence class of ∇ under $Q(G, \mathsf{A})$; thus we will be able to determine the terms v in $cn_{\mathsf{A}}(G)$ if we are able to determine the equations $[v,\nabla]$ in $Q(G, \mathsf{A})$.

If G is the empty set 0 then $Q(0, \mathsf{A})$ is the set $Q(\mathsf{A})$ of *all* T-equations holding in A; hence $Q(\mathsf{A})$ is contained in every $Q(G, \mathsf{A})$. Also, the set $G \times \{\nabla\}$ is contained in $Q(G, \mathsf{A})$ since every h from $H(G, \mathsf{A})$ identifies G and ∇. It follows that $Q(G, \mathsf{A})$ contains

$$R(G, \mathsf{A}) = [Q(\mathsf{A}) \cup (G \times \{\nabla\})] ,$$

the congruence relation *generated by* $Q(\mathsf{A})$ *together with* $G \times \{\nabla\}$ (i.e. the smallest congruence relation on T containing these two sets). Now $R(G, \mathsf{A})$ consists precisely of those T-equations which can be derived, by the calculus of equations *without uses* of the substitution rule, from the equations in $Q(\mathsf{A})$ and in $G \times \{\nabla\}$. As this construction of $R(G, \mathsf{A})$ from $Q(\mathsf{A})$ and $G \times \{\nabla\}$ proceeds by finitary operations, it follows from $[v,w] \in R(G, \mathsf{A})$ that there exists a finite subset F of G such that $[v,w] \in R(F, \mathsf{A})$.

We thus define the consequence operation cn_{A} to be *equational* if $Q(G, \mathsf{A}) = R(G, \mathsf{A})$ holds. In that case, cn_{A} is finitary, the sets $cn_{\mathsf{A}}(G)$ are recursively enumerable in G and $Q(\mathsf{A})$, and the decision problem for cn_{A} becomes equivalent to the word problem (with relations) for the class A .

Theorem 1 cn_{A} *is equational if and only if* $\mathsf{EA} = \mathsf{TA}$ *, i.e. if and only if every algebra from* EA *, generated by an at most countably infinite set, belongs already to* RA *. In that case, the class* EA *is contained in* J *.*

A special case of this theorem was proved in Felscher [65] , sect.4.3 . In order to see that algebras from EA with an uncountable set of generators may not be A-representable, let A consist of the Boolean algebra 2 only such that EA is the class of all Boolean algebras while RA is the class of all Boolean algebras representable as fields of sets. Then RA may be a proper subclass of EA since the prime ideal theorem for Boolean algebras is independent from the usual axioms of set theory. This example shows also that the equational definability of cn_{A} *depends on the size* of the set X of variables of T : cn_{A} may be equational, but an analogous consequence operation cn'_{A} , defined on a term algebra T' with a set X' of larger cardinality, *need not* be equational. Finally, since EJ is not contained in J , the consequence operation cn_{J} is an example of a finitary algebraic consequence which is not equational.

It will simplify the formulation of our proof if we denote by K the class of all algebras in J which are generated by an at most countably infinite set - meaning that they are homomorphic images of T . Now the representation property $\mathsf{K} \cap \mathsf{EA} \subseteq \mathsf{RA}$ is *sufficient* for the equational definability of cn_{A} . For the quotient $T/R(G,\mathbf{A})$ is in K , and it is also in EA since $Q(\mathbf{A}) \subseteq R(G,\mathbf{A})$. Thus $T/R(G,\mathbf{A})$ belongs to $\mathbf{B} = \mathsf{RA}$. Now the classes $\mathsf{A}, \mathsf{SA}, \mathsf{ISA}, \dots, \mathsf{RA}$ all define *the same* consequence operation $cn_{\mathsf{A}} = cn_{\mathbf{B}}$ and also *the same* congruence relations $Q(G,\mathbf{A}) = Q(G,\mathbf{B})$, and it follows from $Q(\mathbf{A}) = Q(\mathbf{B})$ that also $R(G,\mathbf{A}) = R(G,\mathbf{B})$. It thus remains to show $Q(G,\mathbf{B}) = R(G,\mathbf{B})$ which will follow from $Q(G,\mathbf{B}) \subseteq R(G,\mathbf{B})$. But as $T/R(G,\mathbf{B}) = T/R(G,\mathbf{A})$ is in $\mathbf{B}$, the natural homomorphism π from T onto this algebra belongs to $H(G,\mathbf{B})$ whence $Q(G,\mathbf{B}) \subseteq R_{\pi} = R(G,\mathbf{B})$. — The representation property is also *necessary* for the equational definability of cn_{A} . It follows from the definition of $Q(G,\mathbf{A})$ that $T/Q(G,\mathbf{A})$ is isomorphic to a subalgebra of $\Pi < T/R_h \mid h \in H(G,\mathbf{A}) >$, and every T/R_h is in ISA ; thus $T/Q(G,\mathbf{A})$ is in RA . Consider now an algebra B in $\mathsf{K} \cap \mathsf{EA}$; since B is a homomorphic image of T , we may assume that B actually is a quotient T/R with some congruence relation R on T . It follows from $B \in \mathsf{EA}$ that $Q(\mathbf{A})$ is contained in R ; define now

G to be the congruence class of ∇ with respect to R . Then $G \times \{\nabla\} \subseteq R$ implies $R(G, \mathsf{A}) \subseteq R$. Assuming that cn_{A} is equational, we conclude from $R(G, \mathsf{A}) = Q(G, \mathsf{A})$ that $T/R(G, \mathsf{A})$ is in RA ; it thus remains to be shown that $R(G, \mathsf{A})$ is actually equal to R , and this will follow from $R \subseteq R(G, \mathsf{A})$. So let π and π' be the natural homomorphisms from T onto T/R and onto $T/R(G, \mathsf{A})$; the equation $a \supset a = b \supset b = e_A$ still holds in T/R since this algebra belongs to EA . It follows from $[v,w] \in R$, $\pi(v) = \pi(w)$ that $\pi(v \to w) = \pi(v) \supset \pi(w) = \pi(\nabla)$ and also $\pi(w \to v) = \pi(\nabla)$. Thus $v \to w$, $w \to v$ are in G , and now $R_{\pi'} = R(G, \mathsf{A})$ implies $\pi'(v \to w) = \pi'(w \to v) = \pi'(\nabla)$. But $T/R(G, \mathsf{A})$ is in RA and, therefore, in J ; thus the last equalities imply $\pi'(v) = \pi'(w)$ and $[v,w] \in R(G, \mathsf{A})$.

§ 3 CLASSES OF TEST ALGEBRAS

Let A be a subclass of J and let S be the class of test algebras of cn_{A} . Then A is contained in S since cn_{A} is correct for itself; thus $cn_{\mathsf{S}} = cn_{\mathsf{A}}$ since $\mathsf{A} \subseteq \mathsf{S}$ implies $cn_{\mathsf{S}} \subseteq cn_{\mathsf{A}}$. Since $cn_{\mathsf{A}}(0)$ and $cn_{\mathsf{S}}(0)$ determine $Q(0, \mathsf{A}) = Q(\mathsf{A})$ and $Q(0, \mathsf{S}) = Q(\mathsf{S})$, it follows that $Q(\mathsf{A}) = Q(\mathsf{S})$; thus S is contained in EA .

Theorem 2 (i) $\mathsf{S} = \mathsf{TA}$
(ii) *If* cn_{A} *is finitary then* $\mathsf{S} = \mathsf{T_oA}$

Corollary: cn_{A} *is equational if and only if* S *is equational.*

Let us begin the proof by observing that an algebra B in J is A-representable if (and only if) every b in B , distinct from e_B , can be separated from e_B by a homomorphism f from B into an algebra A in A . For if b,b' are distinct elements of B then it follows from $B \in \mathsf{J}$ that either $b \supset b'$ or $b' \supset b$ is distinct from e_B , and if f is a homomorphism satisfying, say, $f(b \supset b') \neq f(e_B)$, $f(b) \supset f(b') \neq e_A$, then $f(b) \neq f(b')$ since also A is in J . — The inclusion $\mathsf{TA} \subseteq \mathsf{S}$ follows already from the general closure properties of S ; for a direct proof assume that C is not in S , and let h be a homomorphism from T into C such that $G = h^{-1}h(\nabla)$ is not cn_{A}-closed. Let B be the image of h in C and let v be in $cn_{\mathsf{A}}(G)$ but not in G . Then h(v) is different from $h(\nabla) = e_C = e_B$ but cannot be separated from e_B by a homomorphism from B into an algebra in A ; thus B is not in RA and C is not in TA . — If cn_{A} is finitary then this proof can be used in order to derive the inclusion $\mathsf{T_oA} \subseteq \mathsf{S}$. For let C,h,G be as before, let v be in $cn_{\mathsf{A}}(G)$ but not in G , and let F be a finite subset of G such that $v \in cn_{\mathsf{A}}(F)$. Let Y be a finite subset of the set X of variables such that v and F belong to the subalgebra T_Y generated by Y in T ; define B to be the image of T_Y under h . Let h' be a homomorphism from T onto B which coincides with h on Y (and may be arbitrary on X-Y) ; thus h' also maps F upon $h(\nabla) = e_C = e_B$ and $h'(v) = h(v)$ is different from e_B. It follows again that h'(v) cannot be separated from e_B by a homomorphism from B into an algebra in A ; thus the finitely generated algebra B is not in RA and C is not in $\mathsf{T_oA}$. — In order to prove the inclusion $\mathsf{S} \subseteq \mathsf{TA}$, consider an algebra C which is not in TA ; let B be a subalgebra of C such that there exists a homomorphism h from T onto B and such that B is not A-representable. Let b be in B such that b, e_B cannot be separated, and let v be a term such that $h(v) = b$. Then v is in $cn_{\mathsf{A}}(h^{-1}(e_B))$ but not in $h^{-1}(e_B)$; hence B is not in S and neither is C. — This concludes the proof of the theorem. The corollary follows immediately from (i) together with theorem 1 .

The definition of the class S depends on the size of the set X of variables generating T . For every infinite cardinal number c , let T_c be a term algebra generated by a set of variables of cardinality c ; let cn^c_{A} be the consequence operation of $u(T_c)$ and let S_c be the corresponding class of test algebras; thus S becomes S_ω. For cardinal numbers c,d with $c < d$ we may assume that T_c is a subalgebra von T_d ; thus the A-filters of T_c are the intersections of $u(T_c)$ with the A-filters of T_d, and a cn^d_{A}-closed subset M of $u(T_d)$ gives rise to a cn^c_{A}-closed subset $M \cap u(T_c)$ of $u(T_c)$. It follows that S_d is a subclass of S_c. Define K_c to be the class of all algebras in J which are homomorphic images of T_c. It follows from the above proof of $\mathsf{S} \subseteq \mathsf{TA}$ that $\mathsf{K}_\omega \cap \mathsf{S}_\omega \subseteq \mathsf{RA} \subseteq \mathsf{S}_\omega$, and since the argument establishing this fact is independent of the size of T , we find $\mathsf{K}_c \cap \mathsf{S}_c \subseteq \mathsf{RA} \subseteq \mathsf{S}_c$ for every cardinal number c . But every algebra in RA belongs, for some sufficiently large c , to $\mathsf{K}_c \cap \mathsf{S}_c$; hence we obtain $\mathsf{RA} = \cap \langle \mathsf{S}_c \mid c \in CARD \rangle$. In the example, mentioned after theorem 1 , the class $\mathsf{S} = \mathsf{S}_\omega$ consists of all Boolean algebras while for larger cardinals c the class S_c may be a proper subclass.

In view of the fact that various classes A will generate the same class TA, it cannot be expected that there is too close a connection between properties of a class A and the consequence cn_{A}. Matters become different, however, for classes A such that $\mathsf{A} = \mathsf{TA}$, i.e. for (full) classes of test algebras, and we shall now investigate particularly the property of cn_{A} of being definable by axioms and rules. In this connection, it will be no restriction to consider only *single valued* rules, for every multiply valued rule can be replaced by a family of single valued ones.

We begin by setting up a calculus cs_o which should describe the consequence cn_J belonging to the *largest* possible class, J itself; consequently, cs_o should be the *weakest* possible calculus. Our axioms and rules are

(γo) $\vdash \; x \to x$,

(γ1) $x \to y \,,\; y \to z \;\vdash\; x \to z$,

(γ2) $x \to x' \,,\; x' \to x \,,\; y \to y' \,,\; y' \to y \;\vdash\; (x \to y) \to (x' \to y')$,

($\gamma 2_f$) $x_o \to x_o' \,,\; x_o' \to x_o \,,\; \ldots \,,\; x_{n-1} \to x_{n-1}' \,,\; x_{n-1}' \to x_{n-1} \;\vdash\; f(x_o,\ldots,x_{n-1}) \to f(x_o',\ldots,x_{n-1}')$,

(γ3) $x \,,\; x \to y \;\vdash\; y$,

(γ4) $x \;\vdash\; y \to x$.

The rule ($\gamma 2_f$) is postulated for every n-ary operation f in Δ which is different from → . — The choice of these axioms and rules is dictated by the criterion for completeness mentioned in § 1 , for we want cs_o to be the weakest calculus satisfying the conditions listed there. So (γo) , (γ1) assure that every relation S(G) is reflexive and transitive, and by (γ2) , ($\gamma 2_f$) it is a congruence relation. Also, (γo) , (γ1) assure that every quotient T/S(G) is partially ordered, and (γ4) says that the terms in cs(G) are mapped upon a largest element under this partial order. Finally, (γ3) assures that every term, mapped upon the largest element of T/S(G) , is in cs(G) , i.e. cs(G) is the congruence class of ∇ .

As cs_o is certainly correct for cn_J , it follows from the criterion for completeness that $cs_o = cn_J$. Thus cn_J can be defined by axioms and rules and, therefore, is finitary. If A is a proper subclass of J then cs_o , although still correct for cn_A , will be complete only for those cn_A which have J as their class of test algebras, i.e. J = TA . In general, however, TA will be a proper subclass of J , and thus a new calculus will be required in order to describe cn_A . This situation is illustrated also by the behaviour of filters: cs_o describes a construction generating J-filters, but A-filters are, in general, larger than J-filters, and so a calculus complete for cn_A will need additional rules in order to generate these larger filters. The formulation of such additional rules will become particularly convenient if, in addition to equations and quasiequations, we employ the syntactical notion of an *inequality*.

Let v,w be terms in T . A homomorphism h from T into an algebra A *satisfies the inequality* $v \mathrel{\dot{\leq}} w$ if $h(v) \leq h(w)$ holds in A ; this is equivalent to

$v \to w \in h^{-1}h(\nabla)$, and $v \to w$ is called the *axiom corresponding to* $v \leq w$. The inequality *holds* or *is valid* in a class A if every homomorphism into an algebra in A satisfies $v \leq w$; this is equivalent to the fact that $v \to w$ belongs to every A-filter. Obviously, an equation holds in A if both the associated inequalities hold in A . An *implication*

(p) if $p_0, \ldots, p_{k-1}$ then q

between equations or inequalities $p_0, \ldots, p_{k-1}, q$ *holds* or *is valid* in A if every homomorphism from T into an algebra in A , satisfying $p_0, \ldots, p_{k-1}$, also satisfies q . Obviously, the validity of an implication (p) with k equations p_j and an equation q , is equivalent to the simultaneous validity of two implications, both of which have the same 2k inequalities p_j', p_j'' to the left and one of the two inequalities q' , q" to the right. If now in (p) the p_j *are* inequalities $v_j \leq w_j$ and if q *is* $v \leq w$ then

(r_p) $v_0 \to w_0, \ldots, v_{k-1} \to w_{k-1} \vdash v \to w$

is called the *rule corresponding to* (p) , and the implication (p) holds in A if and only if every A-filter is closed under (r_p) .

Theorem 3 *Let* B *be a subclass of* J *such that* B = TB , *and let* cs_1 *be a calculus which is complete for* cn_B . *Let* A *be a subclass of* B *which is quasiequational with respect to* B *(i.e.* A *is the intersection of* B *with a quasiequational class)* .

Then we obtain a complete calculus cs *for* cn_A *if we enlarge* cs_1 *by the axioms and rules corresponding to the inequalities, and to the implications between inequalities, of a set of quasiequations defining* A *with respect to* B .

In particular, this theorem may be applied to the case B = J and $cs_1 = cs_0$. — For a proof, observe first that every A-filter is cs-closed for the calculus defined in the theorem; hence cs is correct for cn_A . Let S be the class of test algebras of cs . If B is in S then every {B}-filter in T , being cs-closed, is also cs_1-closed; hence B is also a test algebra of $cs_1 = cn_B$ and, therefore, belongs to B . Since every {B}-filter in T is cs-closed, it now follows from the definition of cs that B also belongs to A . Thus $S \subseteq A$ and, therefore, S = A . Now cs_1 may not explicitly contain the congruence rules $(\gamma 2)$, $(\gamma 2_f)$; however, these rules are *valid for* cs_1 : their conclusions follow by cs_1 from their premisses.

Consequently, they also are valid for cs , and from this it follows that cs is complete for cn_S , i.e. for cn_A .

Theorem 4 *Assume that* $A = TA$. *Then the following statements are equivalent:*

(i) A *is quasiequational* ,
(ii) cn_A *is definable by axioms and finitary rules* ,
(iii) cn_A *is finitary* .

Corollary: Assume, in addition, that the prime ideal theorem for Boolean algebras holds. Then these statements are equivalent to

(iv) A *is elementary (i.e. definable in a first order language with only equations (or inequalities) as atomic formulas).*

Observe first that (i) implies (ii) by theorem 3 and that (ii) implies (iii). If (iii) holds then we define, for every finite set $F = \{v_0,\dots,v_{k-1}\}$ and for every w in $cn_A(F)$, the rule $(\rho_{F,w})$: $v_0,\dots,v_{k-1} \vdash w$; if F is empty then $(\rho_{\emptyset,w})$ is an axiom. It follows that cn_A is the calculus defined by all these axioms and rules; thus (iii) implies (ii). If (ii) holds and if $\vdash w$ is an axiom of cn_A then $\nabla \leq w$ shall be the *corresponding inequality*; if

(r) $v_0,\dots,v_{k-1} \vdash w$

is a rule of cn_A then

(p_r) if $\nabla \leq v_0$, ... , $\nabla \leq v_{k-1}$ then $\nabla \leq w$

shall be the *corresponding implication*. An algebra A has the property that every {A}-filter in T is closed under the axioms and rules (r) of cn_A if and only if the corresponding inequalities and implications (p_r) hold in A . Since A is the class of *all* test algebras of cn_A , it follows that A is the class defined by these inequalities and implications. Thus (ii) implies (i) . As for the corollary, (i) implies (iv) trivially; assume now (iv) . Since A is the class of all test algebras of a calculus, it is closed with respect to subalgebras and direct products, and it also contains all singular algebras. It follows from well known facts in the theory of models (e.g. Malcev [70]) that an elementary class with these properties is quasiequational; the proof depends on the prime ideal theorem for Boolean algebras.

Theorem 5 *A subclass* A *of* J *is quasiequational if and only if* $A = \mathrm{T}A$ *and* cn_A *is finitary .*

Corollary: cn_A *is finitary if and only if* $\mathrm{T}A$ *is quasiequational. In that case,* $\mathrm{T}A$ *is the quasiequational closure* $\mathrm{Q}A$ *of* A .

Let A be quasiequational, defined by inequalities and implications (p) . It follows from theorem 3 that cn_A is defined by the corresponding axioms and rules (r_p) together with the axioms and rules of cs_o . It follows from theorem 4 that the class $\mathrm{T}A$ of test algebras of cn_A is defined by the corresponding inequalities and implications (p_{r_p}) , together with the equations (or inequalities) and implications defining the class J . If $v_i \leq w_i$ is one of the inequalities in (p) then in (p_{r_p}) its place is taken by $\nabla \leq (v_i \to w_i)$. As these inequalities are satisfied by the same homomorphisms, the implications (p) and (p_{r_p}) will hold simultaneously. Thus $\mathrm{T}A$ is defined by the same inequalities and equations as is A and, therefore, is equal to A . — Concerning the corollary, we know that $\mathrm{T}A$ is the class S of test algebras of cn_A , and it follows from $cn_A = cn_S$ that S then is also the class of test algebras of cn_S . Consequently, we have also $\mathrm{TT}A = \mathrm{T}A$ and may apply theorem 5 to $\mathrm{T}A$ in place of A .

There are numerous examples where the construction described in theorem 3 has been applied in order to obtain particular calculi cs ; the case that A is the class of modular ortholattices was treated in Bammert [68] , prop.14 , and the case that A is the class OM of orthomodular lattices was treated in Kalmbach [74] . In particular, Kalmbach has shown that an operation ⊃ may be defined on orthomodular lattices in such a manner that OM may be viewed as a subclass of J and that cn_{OM} may be described as a *pure* MP-calculus. It follows from theorem 3 that now also for every equational subclass A of OM the consequence cn_A may be described as a pure MP-calculus. — Concerning pure MP-calculi in general, the following observation can be made:

Theorem 6 *Assume that* $A = \mathrm{T}A$ *and define* cs *to be the pure* MP*-calculus determined by the (invariant) set* $cn_A(0)$ *of axioms; let* S *be the class of test algebras of* cs .

Then $cs = cn_A$ *holds if and only if* A *is equational and* cs *is complete for* cn_S *. In any case,* $\mathrm{E}A$ *is equal to* S .

Every J-filter F of T is MP-closed; thus F is cs-closed if it contains $cn_A(0)$. Consequently, S consists exactly of those algebras A in J for which every {A}-filter contains $cn_A(0)$. But $cn_A(0)$ is the congruence class of ∇ under $Q(0, A) = Q(A)$, and a filter $h^{-1}h(\nabla)$ is the congruence class of ∇ under R_h; the correspondence between filters and congruence relations now shows that A is in S if and only if $Q(A)$ is contained in R_h for every homomorphism h from T into A. This proves $S = EA$. If A is equational then $S = EA = A$, and if cs is complete for cn_S then $cs = cn_S = cn_A$. Conversely, $cs = cn_A$ implies $cs = cn_S$ by definition of S; thus $cn_A = cn_S$. But A and S both are classes of test algebras and, in that respect, both are maximal for defining cn_A and cn_S; thus A = S. Consequently, $A = EA$ shows that A is equational.

§ 4 DEDUCTIVE CONSEQUENCE OPERATIONS

An MP-map c on the set E of terms is said to be *deductive* if the so-called *deduction rule*

$$\text{if } G, v \vdash w \quad \text{then } G \vdash v \to w$$

is valid for c. If c is finitary and deductive then it follows by repeated application of this rule that the sets c(G) can be computed from c(0) alone by making use of the rule for MP-maps; thus a finitary and deductive MP-map is pure. If c is deductive then all terms of the form

$$(\delta o) \qquad v \to (w \to v)$$

$$(\delta 1) \qquad (u \to (v \to w)) \to ((u \to v) \to (u \to w))$$

are in c(0); conversely, it is known from the elements of propositional logic that a pure MP-map is deductive if all these terms are in c(0). It follows that there exists a *smallest* deductive MP-calculus cd on E, viz. the pure MP-calculus defined by the axioms (δo), (δ1). Consequently, the class D of test algebras of cd consists of all those algebras in J which satisfy the equations corresponding to these axioms. Under the assumption that Δ contains only the operation →, Diego [65], [66] proved that D can be defined by a set DE of equations *without* making references to the class J, namely

$$x \to x = y \to y \quad , \quad (x \to x) \to x = x \quad ,$$

$$x \to (y \to z) = (x \to y) \to (x \to z) \quad , \quad (x \to y) \to ((y \to x) \to x) = (y \to x) \to ((x \to y) \to y) \quad .$$

But the translations between the two characterizations of **D** do not involve arguments which would be affected by the presence of possible further operations; thus they remain in effect also for arbitrary signatures Δ : the class **D** is always equational and is defined by the set DE of Diego's equations. The following theorem collects the more obvious facts about deductive calculi:

Theorem 7 *(A) If* cs *is a finitary and deductive* MP*-calculus then the class* **S** *of test algebras of* cs *is equational. If* cs *is described as the pure* MP*-calculus defined by a subset* M *of* cs(0) *then* **S** *can be defined by the set* DE *together with* $M \times \{\nabla\}$.

(B) If Δ *contains only the operation* $\rightarrow$ *then every deductive* MP*-calculus is complete. If* Δ *is arbitrary, a deductive* MP*-calculus* cs *is complete if and only if, for every additional operation* f *of, say, arity* n *, the set* cs(0) *contains the term* a(f) *which is defined recursively as* a(f,n) *with*

$$a(f,o) = f(x_o,\dots,x_{n-1}) \rightarrow f(x'_o,\dots,x'_{n-1}) \quad ,$$
$$a(f,i) = (x_{n-i} \rightarrow x'_{n-i}) \rightarrow ((x'_{n-i} \rightarrow x_{n-i}) \rightarrow a(f,i-1))$$

for $o < i \leqq n$.

(C) A finitary, deductive MP*-calculus which is complete is also equational.*

It follows from the minimality of cd that cd $\subseteq$ cs for every deductive MP-calculus cs . This means that cs-closed sets are also cd-closed, and thus the class **S** of test algebras of cs is contained in **D** . Consequently, the equations DE hold in **S** , and for every subset M of cs(0) also the equations $M \times \{\nabla\}$ hold in **S** . Consider now an algebra A in which all these equations hold; since **D** is defined by DE , it follows that A belongs to **D** and, therefore, to **J** . Thus every {A}-filter in T is MP-closed, and it also contains M . If cs is assumed to be finitary then it can be described as the pure MP-calculus defined by a suitable subset M of cs(0) ; thus every {A}-filter is cs-closed. Hence A belongs to **S** , and this proves (A) . As for (B), it is well known that, if Δ contains only the operation $\rightarrow$, then cd satisfies the conditions in the criterion for completeness; hence cd $\subseteq$ cs implies that also cs is [deductive and] complete. If Δ is arbitrary, completeness of cs is equivalent to the fact that each of the relations S(G) , determined by cs , is a congruence relation also with respect to the additional operations. It follows with the rule of MP-maps that the axioms a(f) are sufficient for this purpose, and it follows with the deduction rule that they are also necessary. Finally, (C) follows from (A) and theorem 1 .

Let $\mathbf{A}$ be a subclass of $\mathbf{J}$; we shall say that $\mathbf{A}$ *admits principal filters* if, for every A in $\mathbf{A}$ and for every element a of A , the interval $[\![a, e_A]\!]$ (i.e. the set of all x such that $a \leq x \leq e_A$) is an $\mathbf{A}$-filter in A . We denote by $\mathbf{K}$ the class of all algebras in $\mathbf{J}$ which are generated by an at most countably infinite set.

Theorem 8 *Assume that* $\mathbf{A} = \mathsf{T}\mathbf{A}$.

(A) $cn_{\mathbf{A}}$ *is deductive if and only if* $\mathbf{A} \cap \mathbf{K}$ *admits principal filters.*

(B) If $cn_{\mathbf{A}}$ *is finitary then it is deductive if and only if* $\mathbf{A}$ *admits principal filters.*

Let us begin by proving that admissibility of principal filters is a sufficient condition in both (A) , (B) . Assume that $G, v \vdash w$ holds for $cn_{\mathbf{A}}$; it will be shown that, if $G \vdash v \to w$ does *not hold*, then $\mathbf{A} \cap \mathbf{K}$ does *not admit* principal filters. It follows from $\mathbf{A} = \mathsf{T}\mathbf{A}$ that the algebra $A = T/Q(G, \mathbf{A})$ is in $\mathbf{A}$ and, therefore, in $\mathbf{A} \cap \mathbf{K}$; let π be the natural homomorphism from T onto A . It follows from *not* $G \vdash v \to w$ that $\pi(v \to w)$ is not e_A , hence *not* $\pi(v) \leq \pi(w)$, and thus $\pi(w)$ does not belong to $[\![\pi(v), e_A]\!]$. Thus $[\![\pi(v), e_A]\!]$ cannot be an $\mathbf{A}$-filter, for otherwise there would exist a homomorphism f from A into an algebra B in $\mathbf{A}$ such that $f^{-1}(e_B)$ is $[\![\pi(v), e_A]\!]$, and the homomorphism $f\pi$ from T into B would map both G and v to e_B but would not map w to e_B ; this would contradict $G, v \vdash w$. In this argument, the algebra B can be replaced by the image B' of f , and since B' is in $\mathbf{A} \cap \mathbf{K}$ it follows that $[\![\pi(v), e_A]\!]$ is not an $(\mathbf{A} \cap \mathbf{K})$-filter either. — In order to see that in (A) the admissibility is also a necessary condition, let A be in $\mathbf{A} \cap \mathbf{K}$ and assume that A contains an element a such that $[\![a, e_A]\!]$ is not a $(\mathbf{A} \cap \mathbf{K})$-filter; we shall prove that $cn_{\mathbf{A}}$ then is not deductive. Since A is in $\mathbf{K}$, there exists a surjective homomorphism π from T onto A ; let G be $\pi^{-1}(e_A)$ and let v be such that $\pi(v) = a$. Since $\mathbf{A} \cap \mathbf{K}$ contains all singular algebras, there exist $(\mathbf{A} \cap \mathbf{K})$-filters containing $[\![a, e_A]\!]$; since $\mathbf{A}$ is closed with respect to direct products, the intersection F of all these $\mathbf{A}$-filters is an $\mathbf{A}$-filter, and it is even a $(\mathbf{A} \cap \mathbf{K})$-filter since $\mathbf{A}$ is closed with respect to subalgebras. It follows that F must be different from $[\![a, e_A]\!]$, and thus there exists an element b in F such that *not* $a \leq b$. If w is such that $\pi(w) = b$, then it follows from $\pi(v \to w) \neq e_A$ that $G \vdash v \to w$ does not hold for $cn_{\mathbf{A}}$; deductivity will be violated if it can be shown that $G, v \vdash w$ holds. So let h be a homomorphism from T into an algebra B in $\mathbf{A}$ and let h satisfy both G and v ; since $\mathbf{A}$ is closed with respect to subalgebras, we may assume that B is in K . Since $h^{-1}h(\nabla)$ contains $\pi^{-1}(\nabla) = G$ and since π is onto A , it follows that h factors through π : there exists a homomorphism f from A into B such that $h = f\pi$. It follows from $e_B = h(v) = f\pi(v) = f(a)$ that f maps the interval $[\![a,e_A]\!]$ into e_B ; thus $[\![a,e_A]\!]$ is contained in the $(\mathsf{A} \cap \mathsf{K})$-filter $f^{-1}(e_B)$. But F is the smallest $(\mathsf{A} \cap \mathsf{K})$-filter with this property; thus F , and therefore also b , is contained in $f^{-1}(e_B)$. This implies $e_B = f(b) = h(w)$ and proves $G, v \vdash w$.

In order to see that also in (B) the admissibility of principal filters is a necessary condition, some more preparations are needed. It may console the impatient reader that they also prepare the way to some surprising further insights. Consider, for an arbitrary closure operation c on E , the statement

(r) $\qquad v_o,\ldots,v_{n-1} \vdash w$

and let u_o be a variable not occurring in any of the terms in (r) . Then

(r_o) $\qquad u_o \to v_o , \ldots , u_o \to v_{n-1} \vdash u_o \to w$

is called the u_o-*relativization* of (r) . If the rule $(\gamma 4)$ of cs_o is valid for c , then the set c(0) is closed with respect to relativizations of axioms. If c is a deductive MP-map and if (r) is valid for c then also (r_o) is valid for c .

Consider now a quasiequational subclass A of J ; if axioms and finitary rules defining cn_{A} are formulated, it may be assumed that there still exists a variable u_o not occurring in any of them nor in any of the rules $(\gamma o) - (\gamma 4)$ defining $cn_{\mathsf{J}} = cs_o$. Let A be in A ; for every element a of A we define a symmetric relation S_a on u(A) : $[c,c'] \in S_a$ if and only if both $c \supset c'$, $c' \supset c$ are in $[\![a,e_A]\!]$. If $[\![a,e_A]\!]$ is a J-filter then S_a is the corresponding congruence relation.

Lemma 1 *The following two statements are equivalent:*

(do) *for every* a *in* A : S_a *is a congruence relation on* A , *the algebra* A/S_a *is in* J *and* $[\![a,e_A]\!]$ *is a congruence class* .

(do') *every* {A}-*filter on* T *is closed under the* u_o-*relativizations of the rules* $(\gamma 1) - (\gamma 4)$ *of* cs_o .

For assume (do) ; consider the rule $(\gamma 1)$. Let h be a homomorphism from T into A such that $u_o \to (x \to y)$, $u_o \to (y \to z)$ are in $h^{-1}h(\nabla)$; thus $a \leq h(x) \supset h(y)$,

$a \leq h(y) \supset h(z)$ for $a = h(u_o)$. Let π be the natural homomorphism from A onto A/S_a. Since A/S_a is in $\mathcal{J}$, it follows that $\pi h(x) \leq \pi h(y)$, $\pi h(y) \leq \pi h(z)$ and, therefore, $\pi h(x) \leq \pi h(z)$ hold in A/S_a, thus $\pi(h(x) \supset h(z)) = \pi(e_A)$. Since $[\![a,e_A]\!]$ is a congruence class, this implies $a \leq h(x) \supset h(z)$, and so $u_o \to (x \to z)$ is in $h^{-1}h(\nabla)$. Thus $h^{-1}h(\nabla)$ is closed under the u_o-relativization of (γ1), and the other rules are verified analogously. Assume now (do'). In order to see that every S_a is transitive, assume that both $c \supset c'$ and $c' \supset c''$ belong to $[\![a,e_A]\!]$; let h be a homomorphism from T into A sending u_o, x, y, z into a, c, c', c''. Then the terms $u_o \to (x \to y)$, $u_o \to (y \to z)$ are in $h^{-1}h(\nabla)$, and thus $u_o \to (x \to z)$ is in $h^{-1}h(\nabla)$. This implies $a \supset (c \supset c'') = e_A$, and so $c \supset c''$ is in $[\![a,e_A]\!]$. It follows that S_a is transitive and that the $\leq$-relation on A/S_a is transitive as well. The remaining properties stated in (do) are verified analogously.

Lemma 2 *If* (do) *holds then the following statements are equivalent:*

(d1) *for every* a *in* A *:* A/S_a *is in* $\mathcal{A}$

(d1') *every* $\{\mathcal{A}\}$*-filter on* T *is closed under the* u_o*-relativizations of the rules defining* $cn_{\mathcal{A}}$.

That (d1) implies (d1') follows with the same kind of argument which lead from (do) to (do'). Assume now (d1'). Let a be in A ; it has to be shown that every $\{A/S_a\}$-filter on T is closed under the rules defining $cn_{\mathcal{A}}$; let π be the natural homomorphism from A onto A/S_a. Let (r) be one of those rules, let f be a homomorphism from T into A/S_a, and let Y be the finite set of variables occurring in the terms $v_o, \ldots, v_{n-1}, w$ of (r). For every y in Y we choose an element $\eta(y)$ in A such that $\pi\eta(y) = f(y)$; let h be a homomorphism from T into A which maps u_o into a and coincides with η on Y. Then πh and f coincide on the subalgebra of T generated by Y and, therefore, coincide for the terms of (r). Assume now that $v_o, \ldots, v_{n-1}$ are in $f^{-1}f(\nabla)$; then $h(v_o), \ldots, h(v_{n-1})$ are in $\pi^{-1}\pi(e_A)$, i.e. in $[\![a,e_A]\!]$. But then $u_o \to v_o, \ldots, u_o \to v_{n-1}$ are in $h^{-1}h(\nabla)$, and it follows from (d1') that $u_o \to w$ is in $h^{-1}h(\nabla)$. This implies that w is in $f^{-1}f(\nabla)$, and thus $f^{-1}f(\nabla)$ is closed under (r).

An algebra A in $\mathcal{A}$ satisfies (do), (d1) if and only if every interval $[\![a,e_A]\!]$ is an $\mathcal{A}$-filter. Writing $cn^o_{\mathcal{A}}$ instead of $cn_{\mathcal{A}}$, let $cn^1_{\mathcal{A}}$ be the MP-calculus defined by

the axioms and rules of cn^{o}_{A} ,
the u_o-relativizations of the rules (γ1) - (γ4) defining cn_J ,
the u_o-relativizations of the axioms and rules of cn^{o}_{A} .

Let A_1 be the class of test algebras of cn^{1}_{A} . Under the general assumptions made on A , the class A_1 then consists of those algebras A of A in which every interval $[\![a,e_A]\!]$ is an A-filter.

We now can conclude the proof of theorem 8 . For if cn_A is deductive and finitary then all u_o-relativizations of the rules of cn_A and cn_J are valid for cn_A ; hence cn^{1}_{A} coincides with cn^{o}_{A} and, therefore, A_1 coincides with A . In view of the characterization of A_1 , this is what was to be proven.

Consider, again, a quasiequational subclass A of J ; if axioms and finitary rules defining cn_A are formulated, it may be assumed that there still exists a sequence $u_o, u_1, \ldots$ of infinitely many different variables, none of which occurs in these axioms and rules nor in the rules (γ1) - (γ4) . For every natural number i we define an MP-calculus cn^{i+1}_{A} from cn^{i}_{A} making use of u_i-relativizations in the same manner in which cn^{1}_{A} was defined from cn^{o}_{A} . Let cn^{∞}_{A} be the union of the cn^{i}_{A} , let A_i be the class of test algebras of cn^{i}_{A} , and let A_∞ be the intersection of all the A_i . If cs is a deductive MP-calculus such that $cn_A \subseteq cs$ then the u_o-relativizations used for cn^{1}_{A} are valid for cs ; hence also $cn^{1}_{A} \subseteq cs$, and induction shows that $cn^{i}_{A} \subseteq cs$ for every i ; thus $cn^{\infty}_{A} \subseteq cs$, It follows from the definitions that A_∞ is the class of test algebras of cn^{∞}_{A} . But A_∞ has been constructed in such a manner that it admits principal filters; hence cn^{∞}_{A} is deductive. Finally, cn^{∞}_{A} still is finitary, and thus A_∞ is equational. We thus have proved

Theorem 9 *If* A *is quasiequational then* cn^{∞}_{A} *is the smallest deductive* MP-*calculus containing* cn_A *; the class* A_∞ *is its class of test algebras and is equational.*

Obviously, if cn_A happens to be a *pure* MP-calculus then the smallest deductive MP-calculus containing cn_A can be obtained much easier, namely by adjoining the rule (δo) , (δ1) to those of cn_A . Still, for certain, important impure MP-calculi only *one* application of the relativization process will suffice.

Theorem 10 *The smallest deductive* MP-*calculus, being the smallest deductive* MP-*calculus containing* cs_o *, can be obtained from* cs_o *by adjoining the* u_o*-relativizations of the rules of* cs_o .

Let cs_o^1 be the calculus obtained from cs_o by adjoining the u_o-relativizations ($\gamma 10$) - ($\gamma 40$) of the rules ($\gamma 1$) - ($\gamma 4$) . Since the calculus cd is closed under relativizations, it follows from $cs_o \subseteq cd$ that $cs_o^1 \subseteq cd$; for the converse inclusion it will be sufficient to derive (δo) , ($\delta 1$) in cs_o^1 . Now (δo) follows immediately from (γo) and the particular case $x \to x \vdash x \to (y \to x)$ of ($\gamma 40$) . As for ($\delta 1$) , we proceed in four steps:

(I) $\vdash x \to ((x \to y) \to y)$, the *axiom* for modus ponens (MPA) .

Proof:

(1)	$y \to y$	(γo)
(2)	$x \to (y \to y)$	($\gamma 4$) , (1)
(3)	$x \to (x \to (y \to y))$	($\gamma 4$) , (2)
(4)	$x \to ((y \to y) \to x)$	(δo)
(5)	$x \to ((x \to y) \to ((y \to y) \to y))$	($\gamma 20$) , (3) , (4) , (2) , (2)
(6)	$((y \to y) \to y) \to (y \to y)$	($\gamma 4$) , (1)
(7)	$((y \to y) \to y) \to ((y \to y) \to y)$	(γo)
(8)	$((y \to y) \to y) \to y$	($\gamma 30$) , (6) , (7)
(9)	$x \to (((y \to y) \to y) \to y)$	($\gamma 4$) , (8)
(10)	$x \to ((x \to y) \to y)$	($\gamma 10$) , (5) , (9)

(II) $x \to (y \to z) \vdash y \to (x \to z)$, the *rule* for exchange (X)

Proof:

(1)	$y \to ((y \to z) \to z)$	(MPA)
(2)	$x \to (y \to z)$	premiss
(3)	$y \to (x \to (y \to z))$	($\gamma 4$) , (2)
(4)	$y \to (x \to z)$	($\gamma 10$) , (3) , (1)

(III) $\vdash (x \to (y \to z)) \to (y \to (x \to z))$, the *axiom* for exchange (XA)

Proof:

(1)	$y \to ((y \to z) \to z)$	(MPA)
(2)	$z \to (y \to z)$	(δo)
(3)	$y \to (z \to (y \to z))$	($\gamma 4$) , (2)
(4)	$x \to x$	(γo)
(5)	$y \to (x \to x)$	($\gamma 4$) , (4)
(6)	$y \to ((x \to (y \to z)) \to (x \to z))$	($\gamma 20$) , (5), (5) , (1) , (3)
(7)	$(x \to (y \to z)) \to (y \to (x \to z))$	(X) , (6)

(IV) $\vdash (x \to (y \to z)) \to ((x \to y) \to (x \to z))$

Proof:

(1)	$x \to ((x \to (y \to z)) \to (y \to z))$	(MPA)
(2)	$((x \to (y \to z)) \to (y \to z)) \to (y \to ((x \to (y \to z)) \to z))$	(XA)
(3)	$x \to (y \to ((x \to (y \to z)) \to z))$	($\gamma 1$) , (1) , (2)
(4)	$x \to ((x \to y) \to y)$	(MPA)
(5)	$x \to ((x \to y) \to ((x \to (y \to z)) \to z))$	($\gamma 10$) , (4) , (3)
(6)	$(x \to y) \to (x \to ((x \to (y \to z)) \to z))$	(X) , (5)
(7)	$(x \to ((x \to (y \to z)) \to z)) \to ((x \to (y \to z)) \to (x \to z))$	(XA)

(8) $(x \to y) \to ((x \to (y \to z)) \to (x \to z))$ (γ1), (6), (7)
(9) $(x \to (y \to z)) \to ((x \to y) \to (x \to z))$ (X), (8)

For our next considerations we assume that Δ contains (at least) one further operation besides $\to$, and this shall be a binary operation which we denote as $\wedge$. We formulate rules

(αo) $x, y \vdash x \wedge y$

(α1) $x \wedge y \vdash x$

(α2) $x \wedge y \vdash y$

(α3) $x \to x', x' \to x, y \to y', y' \to y \vdash (x \wedge y) \to (x' \wedge y')$

and axioms

(βo) $x \to (y \to (x \wedge y))$

(β1) $(x \wedge y) \to x$

(β2) $(x \wedge y) \to y$

and define three MP-calculi:

cc is obtained from cs_0 by adjoining the rules (αo) - (α3) ;

cc^1 is obtained from cc by adjoining the u_0-relativizations of all defining rules of cc ;

cd_c is obtained from cd by adjoining the axioms (βo) - (β2) .

Observe that, in any MP-calculus, the rules (αo) - (α2) are immediately derivable once (βo) - (β2) have been proved; thus cc appears to be the weakest possible description of conjunction. Oberserve also that cc^1 is obtained from cs_0^1 by adjoining the rules (αo) - (α3) together with their u_0-relativizations (αo0) - (α30) . Observe finally that cd_c is a pure MP-calculus and, therefore, is deductive.

Theorem 11 *(A)* cc^1 *is the smallest deductive* MP-*calculus containing* cc *and* $cc^1 = cd_c$. *The (equational) class of test algebras of* cc^1 *consists of those algebras in* J *which are semilattices for the operation corresponding to* $\wedge$.

(B) For every finitary calculus cm *in which the axioms and rules of* cc *are (at least) derivable, we obtain the smallest deductive calculus containing* cm *by adjoining the* u_0-*relativizations of the axioms and rules defining* cm .

Let us begin the proof by observing that in the MP-calculus cd_c the rules (αo) - (α2) are derivable. Since cd_c is deductive, we obtain (α3) employing the deduction rule to

$$x \wedge y ,\ x \to x' ,\ y \to y' \ \vdash\ x ,\ y ,\ x \to x' ,\ y \to y' \ \vdash\ x', y' \ \vdash\ x' \wedge y' \ .$$

Thus $cc \subseteq cd_c$, and now we conclude $cc^1 \subseteq cd_c$ since cd_c , being deductive, is closed under relativizations. On the other hand, we know $cd = cs_o^1$ from theorem 10 , hence $cd \subseteq cc^1$ since $cs_o^1 \subseteq cc^1$. In order to conclude $cd_c = cc^1$, we have to show $cd_c \subseteq cc^1$, and this will hold if we can prove (βo) - (β2) in cc^1 . Now (β1) follows immediately from (γo) and the particular case $(x \wedge y) \to (x \wedge y) \vdash (x \wedge y) \to x$ of the relativization (α10) , and (β2) follows in the same manner. As for (βo) , we proceed in two steps

(I) $\vdash\ ((x \to y) \wedge (x \to z)) \to (x \to (y \wedge z))$, the *distributive* law (DIS) .

Proof:

(1)	$x \to ((x \to y) \to y)$	(MPA)
(2)	$x \to ((x \to z) \to z)$	(MPA)
(3)	$y \to (x \to y)$	(δo)
(4)	$z \to (x \to z)$	(δo)
(5)	$x \to (y \to (x \to y))$	(γ4) , (3)
(6)	$x \to (z \to (x \to z))$	(γ4) , (4)
(7)	$x \to (((x \to y) \wedge (x \to z)) \to (y \wedge z))$	(α30) , (1) , (5) , (2) , (6)
(8)	$((x \to y) \wedge (x \to z)) \to (x \to (y \wedge z))$	(X) , (7)

(II) $\vdash\ x \to (y \to (x \wedge y))$

Proof:

(1)	$((x \to x) \wedge (x \to y)) \to (x \to (x \wedge y))$	(DIS)
(2)	$x \to x$	(γo)
(3)	$y \to (x \to x)$	(γ4) , (2)
(4)	$y \to (x \to y)$	(δo)
(5)	$y \to ((x \to x) \wedge (x \to y))$	(αo0) , (3) , (4)
(6)	$y \to (x \to (x \wedge y))$	(γ1) , (5) , (1)
(7)	$x \to (y \to (x \wedge y))$	(X) , (6)

This proves the first part of (A) , and the second part is verified immediately. The deduction rule being available, proofs in cc^1 now become quite easy. We can introduce *iterated* conjunctions $\bigwedge\!\!\!\bigwedge \langle v_i \mid i<n \rangle$ for finite sequences of terms and prove derivabilities such as

(κo) $\vdash\ \bigwedge\!\!\!\bigwedge \langle v_i \mid i<n \rangle \to v_i$ for every i such that $i < n$,

(κ1) $\vdash\ (\bigwedge\!\!\!\bigwedge \langle v_i \mid i<n \rangle \to w) \to (v_o \to (v_1 \to \cdots (v_{n-1} \to w) \cdots))$

(κ2) $\vdash\ (v_o \to (v_1 \to \cdots (v_{n-1} \to w) \cdots)) \to (\bigwedge\!\!\!\bigwedge \langle v_i \mid i<n \rangle \to w)$.

Consider now a calculus cm as given in (B) , and let cm^1 be obtained from cm by adjoining the u_o-relativizations of the axioms and rules defining cm . If

$$(r) \qquad v_o, \ldots, v_{n-1} \quad \vdash \quad w$$

is a rule derivable in cm , then we call

$$(a_r) \qquad v_o \to (v_1 \to \cdots (v_{n-1} \to w) \cdots)$$

the *axiom for* (r) . Taking a derivation of (r) in cm , we obtain a derivation of the u_o-relativization $(r|u_o)$ in cm^1 by simply relativizing every single step. In particular, it follows from $cc \subseteq cm$ that $cc^1 \subseteq cm^1$. Define now cd_m to be obtained from cd by adjoining the axioms of cm and the axioms for the rules of cm . In any MP-calculus the axiom (a_r) makes (r) derivable; thus $cm \subseteq cd_m$ and, therefore, also $cm^1 \subseteq cd_m$ holds. If we can show that for every rule (r) of cm the axiom (a_r) is provable in cm^1 , then it will follow from $cd \subseteq cc^1 \subseteq cm^1$ that also $cd_m \subseteq cm^1$; thus $cm^1 = cd_m$ will hold, and cd_m , being a pure MP-calculus, is certainly deductive.

It is only in order to show the provability of (a_r) in cm^1 that we need the presence of the operation $\wedge$ and the calculus cc as an auxiliary, but decisive tool. In the u_o-relativization $(r|u_o)$ of (r) we substitute $\bigwedge\langle v_i | i<n \rangle$ for u_o . It follows from (κo) that the premisses of the result of this substitution are provable in cc^1 and, therefore, in cm^1 . Thus

$$\bigwedge\langle v_i | i<n \rangle \to w$$

becomes provable in cm^1 , and since the rule for MP-calculi is available from cc , it follows with (κ1) that also (a_r) is provable in cm^1 .

REFERENCES

J.Bammert [68] Quasideduktive Systeme und S-Algebren. I / II . Archiv f.Math. Logik und Grundlagenforsch. 11 (1968) 56-67 and 101-112 .

W.J.Blok , P.Köhler , D.Pigozzi [82] Algebraic Logic. Manuscript .

J.Czelakowski [81] Equivalential Logics. I / II . Studia Logica 40 (1981) 227-236 and 355-372 .

A.Diego [65] Sobre algebras de Hilbert. Bahia Blanca 1965

A.Diego [66] Sur les algèbres de Hilbert. Paris 1966

W.Felscher [65] Adjungierte Funktoren und primitive Klassen. Sitzungsber. Heidelberger Akad.Wiss., Math.-nat.Kl. 1965, 445-509

G.Kalmbach [74] Orthomodular Logic. Zeitschr.f.Math.Logik und Grundlag. d.Math. 20 (1974) 395-406

C.H.Lambros [79] A generalized theorem concerning a restricted rule of substitution in the field of propositional calculi. Notre Dame J. of Formal Logic 20 (1979) 760-764

J.Łos, R.Suszko [58] Remarks on sentential logics. Indagat.Math. 20 (1958) 177-183

A.I.Malcev [70] Algebraičeskie Sistemi. Moskva 1970

H.Rasiowa [74] An algebraic approach to non-classical logics. Amsterdam 1974

H.Rasiowa, R.Sikorski [53] Algebraic treatment of the notion of satisfiability. Fundamenta Math. 40 (1953) 62-95

B.Sobociński [74] A theorem concerning a restricted rule of substitution in the field of propositional calculi. I. Notre Dame J. of Formal Logic 15 (1974) 465-476

Walter Felscher
Alte Steige 10
Obernau / Neckar

Jürgen Schulte Mönting
Mathematisches Institut
Universität Tübingen

Universal Algebra and its Links with Logic,
Algebra, Combinatorics, and Computer Science
Proc. "25. Arbeitstagung über Allgemeine Algebra", Darmstadt 1983
P. Burmeister et al. (eds.)

67 – 77

ON CONSTRUCTING MANY NON-ISOMORPHIC ALGEBRAS

W. Hodges

In the early 1970s, model theorists proved several results of the form 'Class K contains 2^λ pairwise non-isomorphic structures of cardinality λ'. Saharon Shelah, who almost has a monopoly on these theorems (see Chapter VIII of [7]), argued in [8] that results of this type should be thought of as <u>non-structure theorems</u>. Putting it roughly, they show that there is no way of classifying all the structures in K up to isomorphism by a small family of cardinal invariants. There are too many isomorphism types to go round.

This is a valuable perspective, but it seems a pity to stop there. For example a group theorist, faced with 2^λ non-isomorphic examples of his favourite kind of group of cardinality λ, is likely to be curious about how different from each other they are. Any of the following would be more use to him than a bare existence theorem:

(1) There are 2^λ groups G_i $(i < 2^\lambda)$ of the relevant kind, all of cardinality λ, such that if $i \neq j$ then G_i is not embeddable in G_j.

(2) There are 2^λ groups ..., such that if $i \neq j$ then any group embeddable both in G_i and in G_j is 'small'.

(3) There are 2^λ groups ..., such that if $i \neq j$ then there is a 'small' group which is embeddable in G_i but not in G_j.

With a suitable interpretation of 'small', (2) and (3) each imply (1).

Shelah himself has usually published his results in strongest form, so that in many applications we have at least the analogue of (1). Other authors have not always pressed so hard. In this note I examine two cases where the published results can be improved by reexamining known proofs. The first case concerns boolean powers, and the second revolves around a construction which Keisler used to build many models of cardinality ω_1. I have also taken the liberty of including an unpublished result of Shelah which refines Keisler's construction when the continuum hypothesis holds.

My thanks to John Baldwin and Saharon Shelah for some helpful correspondence, and to the organisers of the 25th Arbeitstagung at Darmstadt.

1. Boolean powers

Boolean powers $A[B]^*$ will be understood as in Chapter IV §5 of Burris and Sankappanavar [3]. In particular they are bounded. If A has more than one element and B is infinite, then $|A[B]^*| = |A| + |B|$.

John Baldwin and Ralph McKenzie in a recent paper [1] give a clever argument to prove that any countable structure A not satisfying a certain condition (*) (they call it (A)) has 2^λ pairwise non-isomorphic boolean powers of cardinality λ for each $\lambda > \omega$. The condition (*) says:

(*) For every $n < \omega$ and every atomic formula $\psi(\bar{x},\bar{y})$ with n+n variables, the sentence $\forall\bar{x}\bar{y}(\psi(\bar{x},\bar{x})\wedge\psi(\bar{y},\bar{y})\wedge\psi(\bar{x},\bar{y}) \rightarrow \psi(\bar{y},\bar{x}))$ holds.

They show that if an algebra A is non-abelian (i.e. if the centre of A as defined on p. 83 of [3] is not A^2), then A fails to satisfy (*). Hence for $\lambda > \omega$ and countable languages their result generalises the result of Burris [2], that a non-trivial congruence-distributive variety V contains 2^λ algebras in each cardinality $\lambda \geq |\text{language of } V|$, all of them boolean powers of a simple algebra.

They draw the corollary that if T is a universal Horn theory with a countable model not satisfying (*), then T has 2^λ non-isomorphic models in every cardinality $\lambda \geq \omega$. (For $\lambda = \omega$ there is a special argument.) They point out that if A fails to satisfy (*), then infinite powers of A have unstable first-order theories in the sense of Shelah [7], so that for uncountable λ the corollary follows also from Shelah's non-structure theorem for unstable theories. They note that the proof by the Shelah route doesn't make the structures boolean powers, so that it gives a weaker result.

Here they miss a point. The Shelah proof is in one significant way stronger than the boolean power proof. When λ is regular, Shelah's proof finds the 2^λ structures so that none is embeddable in any of the others, just as in (1) above. This follows at once from the fact that the unstable formula given by the failure of (*) is quantifier-free, and quantifier-free formulas are preserved by embeddings.

One can do better still. For regular λ, Shelah's argument plus a little chewing-gum <u>does</u> give boolean powers, <u>and</u> they can be chosen to be pairwise non-embeddable. The Baldwin-McKenzie argument gives no information about embeddings.

THEOREM 1. Let λ be a regular uncountable cardinal and A a structure of cardinality $< \lambda$, in which (*) fails. Then there is a family of 2^λ boolean powers $A[B_i]^*$ $(i < 2^\lambda)$ of A, all of cardinality λ, such that if $i \neq j$ then $A[B_i]^*$ is not embeddable in $A[B_j]^*$.

<u>Proof</u>. For technical convenience we assume A has an element named by a constant ∞. If not, then expand A to a structure A^+ by adding a new element ∞ and a 1-ary relation symbol P picking out the set of old elements. Then for each boolean algebra B, $A[B]^*$ is the relativisation of $A^+[B]^*$ to P, and the arguments below can be adapted to build the structures $A^+[B_i]^*$ so that there are no embeddings between their P-parts.

Also for technical convenience we assume that A has elements d_0, d_1 such that for some atomic formula ψ,

$$A \models \psi(d_0,d_0) \wedge \psi(d_1,d_1) \wedge \psi(d_0,d_1) \wedge \neg\psi(d_1,d_0).$$

Since (*) fails, there are some such n-tuples $\bar{d}_0$, $\bar{d}_1$. To reduce to single elements one can replace A by a two-sorted structure (A,A^n) with a natural map from A to A^n and n projection functions in the other direction. It is important that these new functions are defined by universal Horn sentences, so that the passage from A to (A,A^n) commutes with taking boolean powers. But for simplicity I assume below that A is one-sorted.

With these assumptions, we construct two two-sorted first-order theories T_1 and T_2. T_1 will be designed so that:

(4) If B is any boolean algebra then T_1 has a model of form $(B,A[B]^*,\bar{F})$ where B, $A[B]^*$ are respectively the structures of first sort and second sort, $\bar{F}$ is a family of functions relating the two sorts, and every element of $A[B]^*$ is the value of a function in $\bar{F}$ applied to some elements of B. If $(B,C,\bar{G})$ is also a model of T_1 such that every element of C comes from B by $\bar{G}$, then $C \cong A[B]^*$.

To construct T_1, introduce for each positive integer n and each ordered n-tuple $\bar{a} = (a_1,\ldots,a_n)$ of elements of A an n-ary function symbol $f_{\bar{a}}$ from the first sort to the second. $\bar{F}$ in (4) will consist of the interpretations of these symbols $f_{\bar{a}}$. For any ordered n-tuple $\bar{b} = (b_1,\ldots,b_n)$ of elements of a boolean algebra B, the intended interpretation of $f_{\bar{a}}(\bar{b})$ is:

If $\bar{b}$ is a partition of 1 by nonzero elements, then $f_{\bar{a}}(\bar{b})$ is the element c of $A[B]^*$ such that $c(x) = a_i$ whenever x is an ultrafilter on B containing b_i $(1 \le i \le n)$. If $\bar{b}$ is not a partition of 1 by nonzero elements, then $f_{\bar{a}}(\bar{b})$ is ∞.

T_1 will contain, first, some sentences to control when $f_{\bar{a}}(\bar{x}) = f_{\bar{a}'}(\bar{y})$ holds. For example if π is a permutation of $\{1,\ldots,n\}$ and $\bar{a}$ has no repetitions, then T_1 will contain (using an obvious shorthand):

$$\forall\bar{x}\bar{y}(\bar{x},\ \bar{y} \text{ each partition } 1 \rightarrow (f_{\bar{a}}(\bar{x}) = f_{\pi\bar{a}}(\bar{y}) \leftrightarrow \bigwedge_{1\le i\le n} y_i = x_{\pi(i)})).$$

Second, T_1 will contain sentences to specify the diagram of $A[B]^*$. For example if R is a 2-ary relation symbol of the language of A, then for all n-tuples $\bar{a}$, $\bar{a}'$

in A, T_1 will contain:

$$\forall\bar{x}(\bar{x} \text{ partitions } 1 \rightarrow R(f_{\bar{a}}(\bar{x}),f_{\bar{a}'}(\bar{x}))) \qquad \text{(if for all } i,\ A \models R(a_i,a_i'));$$

$$\forall\bar{x}(\bar{x} \text{ partitions } 1 \rightarrow \neg R(f_{\bar{a}}(\bar{x}),f_{\bar{a}'}(\bar{x}))) \qquad \text{(otherwise).}$$

This completes the definition of T_1.

Let T_2 be the union of T_1 and the Skolemised theory of some infinite boolean algebra in the language of the first sort. T_2 has a model whose first sort contains an infinite indiscernible sequence η of elements, ordered by the ordering relation $<$ of the boolean algebra. Let Φ be the set of first-order formulas $\phi(v_0,v_1,\ldots)$ satisfied by increasing tuples from η.

Write $x \prec y$ for: $x < y \vee (\psi(x,y) \wedge \neg\psi(y,x))$. By the proof of Theorem VIII.2.2(1) in Shelah [7], together with (4) above, there are a family of 2^λ linear orderings η_i $(i < 2^\lambda)$ of cardinality λ, and a family $C_i = (B_i, A[B_i]^*, \bar{F}_i)$ $(i < 2^\lambda)$ of models of T_2, such that the following holds:

(5) For each i, η_i is a set of elements of B_i which generates C_i, increasing tuples from η_i satisfy Φ, and if $i \neq j$ then no subset of C_i is ordered in order-type η_j by $\prec$.

In particular if $i \neq j$ then no subset of $A[B_i]^*$ is ordered in order-type η_j by ψ.

To reach the theorem, we need only show that some subset of $A[B_i]^*$ is ordered in order-type η_i by ψ. This will use the definition of the boolean power $A[B_i]^*$. By (5), η_i is a subset of B_i ordered by $<$. For each b in η_i, let c_b be the element of $A[B_i]^*$ such that for each ultrafilter x on B_i,

$$c_b(x) = \begin{cases} d_1 & \text{if } b \in x, \\ d_0 & \text{otherwise.} \end{cases}$$

Suppose now that $b < b'$ in η_i. Then c_b and $c_{b'}$ agree on all ultrafilters x except those containing b' but not b, and for these x we have

$$A \models \psi(c_b(x), c_{b'}(x)) \wedge \neg\psi(c_{b'}(x), c_b(x)).$$

It follows that $A[B_i]^* \models \psi(c_b, c_{b'}) \wedge \neg\psi(c_{b'}, c_b)$. Hence ψ orders $\langle c_b : b \in \eta_i \rangle$ in order-type η_i. □

Let me add three remarks. First, we put no Skolem functions on $A[B_i]^*$, so that the usual argument to show $C_i \equiv C_j$ fails here. But there are Skolem functions on the boolean algebras B_i, and hence $B_i \equiv B_j$ for each $i < j$. It follows that $A[B_i]^* \equiv A[B_j]^*$ by a Feferman-Vaught argument (cf. Theorem 4.3 (i) in Burris [2]). Thus if B is any infinite boolean algebra, we can choose all the $A[B_i]^*$ to be elementarily equivalent to $A[B]^*$.

Second, this technique of Shelah only works when λ is uncountable. So for $\lambda = \omega$ the argument of Baldwin and McKenzie has no competitors. Observe that if A is a countable atomless boolean algebra then so is $A[B]^*$ for each countable boolean algebra B, so that Theorem 1 is simply false when $\lambda = \omega$.

Third, the argument of Theorem 1 is independent of the cardinality of A; the Baldwin-McKenzie argument seems to make essential use of the countability of A. However, λ must be regular (or a singular cardinal such that $2^\lambda = 2^\mu$ for some $\mu < \lambda$). Shelah [9] has methods for constructing many non-isomorphic models in singular cardinalities, but it was not clear to me that they can be used to get boolean powers.

2. Keisler's construction in cardinality ω_1

In §5 of [5] and Chapter 31 of [6], H. J. Keisler proved a theorem equivalent to the following:

Let $L' \subseteq L$ be countable first-order languages and ϕ a sentence of $L_{\omega_1\omega}$. Suppose ϕ has a model in which uncountably many complete L'-types are realised. Then there is a family B_i $(i < 2^{\omega_1})$ of models of ϕ, all of cardinality ω_1, such that if $i \neq j$ then some L'-type is realised in B_i but not in B_j.

The conclusion implies that if $i \neq j$ then $B_i|L'$, the L'-reduct of B_i, is not elementarily embeddable in $B_j|L'$.

Exactly the same proof works if we replace 'complete L'-type' by 'complete quantifier-free L'-type', where a complete quantifier-free L'-type is a set of the form $\{\psi(\bar{v}) : \psi$ quantifier-free in L' and $A \models \psi(\bar{a})\}$ for some structure A and elements $\bar{a}$. Taking L' = L, the complete quantifier-free L-types realised in a structure A are for practical purposes just the isomorphism types of finitely generated substructures of A. Thus we have:

THEOREM 2 (Keisler, lightly glossed). Let ϕ be a sentence of $L_{\omega_1\omega}$, and suppose ϕ has a model with uncountably many isomorphism types of finitely generated substructures. Then there is a family B_i $(i < 2^{\omega_1})$ of models of ϕ, all of cardinality ω_1, such that if $i \neq j$ then some finitely generated structure is embeddable in B_i but not in B_j. □

COROLLARY 3. Let V be a variety in a countable language, and suppose there is an algebra in V with uncountably many pairwise non-isomorphic finitely generated subalgebras. Then there is a family B_i $(i < 2^{\omega_1})$ of 2^{ω_1} existentially closed V-algebras of cardinality ω_1, such that for any $i \neq j$ there is some finitely generated algebra embeddable in B_i but not in B_j.

Proof. The existentially closed V-algebras are those which omit a certain countable family of types, and hence they are defined by a sentence of $L_{\omega_1\omega}$. (Cf. Hirschfeld and Wheeler [4] Proposition 6.1.) □

Martin Ziegler derived this corollary in the case of groups, using a considerable amount of ingenious group theory and recursion theory ([10] Folgerung p. 508). But evidently Corollary 3 applies equally well to many other varieties.

The types in Theorem 2 and Corollary 3 are over the empty set of parameters. What can one say in terms of types over sets of parameters?

Corollary 4 below is folklore. Theorem 5 is an unpublished result of Shelah, which he kindly sent me in a letter dated October 1981. 'Type' means 'complete quantifier-free type' throughout; like Keisler we could also have restricted the types to a sublanguage L' of L.

COROLLARY 4. Assume $2^{\omega} < 2^{\omega_1}$. Suppose L is a countable first-order language, and ϕ is a sentence of $L_{\omega_1\omega}$ which has a model in which uncountably many types over some countable set are realised. Then ϕ has 2^{ω_1} pairwise non-isomorphic models of cardinality ω_1.

Proof. There are only $\omega_1^{\omega} = 2^{\omega} < 2^{\omega_1}$ distinct ways of adding countably many constants to a structure of cardinality ω_1. □

THEOREM 5. Assume $2^{\omega} = \omega_1$. Suppose L is a countable first-order language, and ϕ is a sentence of $L_{\omega_1\omega}$ which has a model in which uncountably many types over some countable set are realised. Then ϕ has a family B_i $(i < 2^{\omega_1})$ of models of cardinality ω_1, such that if $i \neq j$ then B_i is not embeddable in B_j.

Proof. Recall Keisler's argument in [5], [6]. There is a countable fragment L^+ containing both L and a sentence ψ which implies ϕ and ensures that 'uncountably many types are realised'. For each $s \in {}^{(<\omega_1)}2$ he finds a countable model A_s of ψ, a countable set Π_s of types (here, complete quantifier-free L-types over the empty set) and a type p_s such that for all s, t $\in {}^{(<\omega_1)}2$,

(6) A_s has an uncountable L^+-elementary extension;

(7) if $s \subseteq t$ then $A_s \preccurlyeq_{L^+} A_t$ and $\Pi_s \subseteq \Pi_t$; if t is of limit length then $A_t = \bigcup_{s\subset t} A_s$ and $\Pi_t = \bigcup_{s\subset t} \Pi_s$;

(8) A_s realises p_s and strongly omits all the types in Π_s;

(9) $p_{s0} \in \Pi_{s1}$ and $p_{s1} \in \Pi_{s0}$.

The definitions go by induction on the length of s. For each $\eta \in {}^{\omega_1}2$ he puts $B_\eta = \bigcup_{s\subset\eta} A_s | L$. If $s0 \subset \eta$ and $s1 \subset \zeta$ then p_{s0} is realised in B_η but not in B_ζ, and vice versa with p_{s1}.

We slightly adapt this format. First we add a book-keeping clause:

(10) For each s the set of elements of A_s is a countable ordinal.

Adding to L^+ new constants for the elements of ω, we rewrite ψ so that it ensures 'uncountably many types over ω are realised'. The sets Π_s will be countable sets of types over countable subsets of A_s. Otherwise (6)-(8) remain as before. We delete (9) and proceed as follows.

Since $2^\omega = \omega_1$, the elements of ${}^{(<\omega_1)}2$ can be listed without repetition as s_i $(i < \omega_1)$ so that if $s_i \subset s_j$ then $i < j$. We write $s_i \prec s_j$ iff $i < j$. Also the functions $f\colon \omega \to \omega_1$ can be listed as f_i $(i < \omega_1)$. If q is a type over ω, we write fq for the corresponding type over $f\omega$. We define A_{s_i}, Π_{s_i} and p_{s_i} by induction on i, so that (6)-(8), (10) hold and:

(11) p_{s_i} is distinct from all q such that for some $j, k < i$, $f_j\omega \subseteq A_{s_k}$ and f_jq is not strongly omitted in A_{s_k} (there are only countably many such q, by [5] Lemma 5.3).

(12) If s_i is s_k0 or s_k1, then Π_{s_i} contains all types f_jq such that $j < i$, $f_j\omega \subseteq A_{s_k}$, q is realised in some A_s $(s \prec s_i)$ and f_jq is strongly omitted in A_{s_k}.

The structures B_η $(\eta \in {}^{\omega_1}2)$ are defined as before.

Suppose now for contradiction that $\eta \neq \zeta$ and $g\colon B_\eta \to B_\zeta$ is an embedding. Then $g|\omega = f_j$ for some j. Choose $i > j$ so that

(13) $s_i \subset \eta$, $s_i \not\subset \zeta$, and there is $\zeta|m \prec s_i$ such that $f_j\omega \subseteq A_{\zeta|m}$,

and consider $p = p_{s_i}$. By (8), B_η realises p. We shall reach a contradiction by showing that $gp = f_jp$ is not realised in B_ζ. It suffices to show that if $f_j\omega \subseteq A_{\zeta|m}$ then f_jp is strongly omitted in $A_{\zeta|m}$, by induction on m.

When $\zeta|m \prec s_i$ (a case which does occur, by (13)), then (11) implies that $f_j p$ is strongly omitted in $A_{\zeta|m}$. Next, let m be the first ordinal such that $\zeta|m+1 \not\prec s_i$. Then $\zeta|m+1 \neq s_i$ by (13), and hence $s_i \prec \zeta|m+1$. If $\zeta|m \prec s_i$ then $A_{\zeta|m}$ strongly omits $f_j p$ as above. If $\zeta|m \not\prec s_i$ then m is a limit ordinal, and $A_{\zeta|m}$ strongly omits $f_j p$ since $A_{\zeta|k}$ strongly omits $f_j p$ for all $k < m$. Either way, $A_{\zeta|m}$ strongly omits $f_j p$ and hence $f_j p \in \Pi_{\zeta|m+1}$ by (12). It follows by (7), (8) that B_ζ omits $f_j p$. □

REFERENCES

[1] John T. Baldwin and Ralph N. McKenzie, Counting models in universal Horn classes, Algebra Universalis

[2] Stanley Burris, Boolean powers, Algebra Universalis 5 (1975) 341-360.

[3] Stanley Burris and H. P. Sankappanavar, A course in universal algebra, Springer, New York 1981.

[4] Joram Hirschfeld and William H. Wheeler, Forcing, arithmetic, division rings, Lecture Notes in Math. 454, Springer, Berlin 1975.

[5] H. Jerome Keisler, Logic with the quantifier "There exist uncountably many", Ann. Math. Logic 1 (1970) 1-93.

[6] H. Jerome Keisler, Model theory for infinitary logic, North-Holland, Amsterdam 1971.

[7] Saharon Shelah, Classification theory and the number of non-isomorphic models, North-Holland, Amsterdam 1978.

[8] Saharon Shelah, The lazy model-theoretician's guide to stability, Logique et Analyse (1975) 241-308; reprinted in Six days of model theory, ed. P. Henrard, Editions Castella, Albeuve, Switzerland 1977, pp. 9-76.

[9] Saharon Shelah, The number of non-isomorphic models of an unstable first-order theory, Israel J. Math. 9 (1971) 473-487.

[10] Martin Ziegler, Algebraisch abgeschlossene Gruppen, in Word problems II, the Oxford book, ed. S. I. Adian and others, North-Holland, Amsterdam 1980, pp. 449-576.

Wilfrid Hodges
Department of Mathematics
Bedford College
Regent's Park
London NW1 4NS

POSTSCRIPT

After writing the paper I learned that Shelah already knew Theorem 1. His proof is less direct than the one above: it involves expressing the class of boolean powers of A as the class of reducts to L(A) of models of a sentence of $L_{\lambda^+\omega}$ where $\lambda = |A| + |L(A)|$, and then invoking some infinitary model theory (as in his paper 'A combinatorial problem; stability and order for models and theories in infinitary languages', Pacific J. Math. 41 (1972) 247-261). This approach finds 2^λ boolean powers of cardinality λ for singular λ as well as regular.

I am a little disappointed, because I meant Theorem 1 to illustrate how Shelah's techniques for constructing many pairwise non-embeddable models can be used even when the class of structures is not easily definable by a theory. I hope it makes this point anyway.

A forthcoming paper of Shelah ('Construction of many complicated uncountable structures and boolean algebras', Israel J. Math.) contains more information about how to get many pairwise non-embeddable models.

Universal Algebra and its Links with Logic,
Algebra, Combinatorics, and Computer Science
Proc. "25. Arbeitstagung über Allgemeine Algebra", Darmstadt 1983
P. Burmeister et al. (eds.)

78 – 84

SOME ASPECTS OF NONSTANDARD METHODS IN GENERAL ALGEBRA

M. M. Richter

§1 Introductory Remarks

Nonstandard mathematics is in some sense the theory of ideal points. Ideal points are introduced in various ways.
Nonstandard methods are concerned with infinitary situations (not like the introduction of negative numbers). Ideal points are used to "make infinitary objects finite".
In some sense the basic paradox of nonstandard analysis is the extension of infinite sets to finite sets. This is familiar to topologists: With some justification we have the equality compact = finite; and compactification is an extension. Therefore the nonstandard methods have been mainly used in topology and analysis. This does not mean that they are superfluous in algebra: topological arguments, completions, limits etc. are familiar in algebra whenever we deal with infinite situations. Traditionally, nonstandard arguments occur in the form of explicit constructions like ultraproducts or ultralimits in order to describe a certain phenomenon, example or counterexample. It seems useful, however, to describe this method in a more abstract way. We find it convenient to use the axiomatic approach rather than explicit constructions. From the many applications in this area we also select two special ones which are connected with two typical constructions in universal algebra, namely inverse and direct limits. We also are more concerned with motivations and examples rather than giving detailed technical proofs which are omitted or briefly sketched.

§2 Axiomatic nonstandard mathematics and some remarks on monads.

The axioms of internal set theory describe the nonstandard set world as a whole.
We have as undefined predicates "=", "ε" and a new predicate standard(x). Formulas without "standard(x)" are internal; external formulas contain the standard predicate. The quantifiers $(\forall^{st}x)\varphi$

and $(\forall^{stfin}x)\varphi$ mean $\forall x(\text{standard}(x) \to \varphi)$ and $(\forall^{st}x)(x \text{ finite} \to \varphi)$.

Nelson's axioms for internal set theory are:

Axiom (O) ZFC holds for "ε" and "=".

Axiom (T) Transfer:

For internal P with the variables $x, y_1, \ldots, y_n$:

$$\forall^{st}\, y_1 \ldots y_n\ [\forall^{st}x\ P(x,y_1,\ldots,y_n) \leftrightarrow \forall x\ P(x,y_1,\ldots,y_n)]$$

Axiom (I) Ideal point:

For internal P:

$$[\forall^{stfin}\ x\ \exists y\ \forall z \in x\ (P(y,z))] \leftrightarrow [\exists y\ \forall^{st}\ z\ (P(y,z))]$$

Axiom (S) Standard sets:

For internal or external P with x as a free variable:

$$\forall^{st}\ y\ \exists z^{st}\ \forall^{st}\ x\ [x \in z \leftrightarrow (x \in y \wedge P(x))]$$

Sets of the form $\{x \in X \mid \varphi(x)\}$ with φ external are not in the internal world but in the naive universe of sets, they are called external. Some very important external sets are the monads. Let X be standard and $\mathcal{F}$ a standard filter on X.

Def.: The monad $\mu_{\mathcal{F}}$ of $\mathcal{F}$ is the external set
$\mu_{\mathcal{F}} = \cap\ (Y \mid Y \in \mathcal{F},\ Y \text{ standard})$.

Equivalently:

Take for $\varphi(x)$ the formula

$$(\forall^{st}\ Y \in \mathcal{F})\ \ (x \in Y),$$

which is an abbreviation for $(\forall Y \in \mathcal{F})\ (\neg\text{standard } (Y) \vee x \in Y)$, then $\mu_{\mathcal{F}} = \{x \in X \mid \varphi(x)\})$.

Here the standard predicate occurs only negatively (i. e. inside of an odd number of negation symbols). Such formulas are called topological formulas (or $\forall^{st}$ - formulas). These formulas characterize monads:

Theorem on monads:

An external subset $A \subseteq X$, X standard, is the monad of some standard filter $\mathcal{F}$ on X

iff

$A = \{x \in X \mid \varphi(x)\}$ for some topological formula φ.
The filter $\mathscr{F}$ is uniquely determined, its standard elements are the standard $Y \supseteq A$.

§3 Inverse limits:

One important use of inverse limits in general algebra is to extend properties of finite structures to some infinite structures, say from finite to profinite groups. This transfer is also (and more systematically) obtained by using nonstandard finite models. Profinite structures carry a natural compact uniformity the monad of which usually is very simple to describe in the nonstandard model. We will consider some examples. X denotes a standard set.

a) Define $\approx \subseteq \mathcal{P}(X \times X)$ by

$$Y \approx Z \Longleftrightarrow (\forall^{st} x \in X)\ (x \in Y \leftrightarrow x \in Z).$$

It is a monad on $\mathcal{P}(X \times X)$; because it is an equivalence relation it is a uniformity; because each point (=set) is nearstandard (i.e. in the monad of some standard point) it is compact. A base of the uniformity is given by sets of the form

$$B_{x_1,\ldots,x_n} = \{(Y,Z) \mid x_i \in Y \leftrightarrow x_i \in Z,\ 1 \le i \le n\},$$

this means $\mathcal{P}(X) \cong 2^X$ has the product topology of the discrete spaces $2 = \{0,1\}$. If one restricts the topology to some $\mathfrak{X}' \subseteq \mathcal{P}(X)$ then the trace topology is in general no longer compact; very often one looks for the limits in the greater space $\mathcal{P}(X)$ (in this topology).

b) Take a standard finite alphabet $\Sigma = \{a_1,\ldots,a_n\}$, $X = \Sigma^*$ the free monoid. On $\mathcal{P}(X)$ one has the operations $Y \cup Z$, $Y \cdot Z$ (concatenation), Y^n, Y^* where $Y^* = \bigcup_{n=0}^{\infty} Y^n$.

The operations "$\cup$" and "$\cdot$" are continuous in the uniformity defined by "$\approx$". If $\mathfrak{A}$ is the subalgebra von $(\mathcal{P}(X),\cup,\cdot)$ generated by the singletons its completion is the algebra of regular events; the "$*$" - operation is obtained from some nonstandard power Y^ω. Model - theoretically one can obtain the completion as one gets the reals from the rationals via nonstandard methods, namely through external factorizations by a monad.

From the viewpoint of automata theory the "$*$" corresponds to a loop in the automaton; this means that automata with loops are "limits of loop - free automata". If we regard flow charts as automata this phenomenon was descibed by D. Scott in his work on continuous lattices.

c) Computable functions (for simplicity on $\mathbb{N}$):
Define "$\approx$" by

$$f \approx g :\Longleftrightarrow (\forall^{st} n \in \mathbb{N})(\forall^{st} t \in \mathbb{N}) \ (f_t(n) = g_t(n)).$$

Here $f_n(t)$ is the value at the time t of the excution sequence of f for input n.

We obtain: Up to "$\approx$" each standard partial recursive function is primitive recursive; here the limit points of primitive recursive functions are in general not primitive recursive. This example is very similar to the previous one:
It can be extended to define the semantics of programming languages. The technique can be used for metaproofs concerning verification calculi for proving the correctness of programs.

d) Take $F \supseteq E$ a standard Galois extension of fields and let be

$$G = \mathrm{Gal}(F/E) \text{ its Galois group.}$$

For $\sigma, \tau \in G$ put

$$\sigma \approx \tau :\Longleftrightarrow (\forall^{st} x \in F)(\sigma(x) = \tau(x)).$$

The topology defined this way is the Krull topology.
Because each automorphism maps standard elements to standard elements and the property of being "automorphism" depends only on the behaviour of finitely many arguments we see that the Krull topology is compact.

e) We generalize the last two examples. Take a standard directed inverse system

$$(X_\alpha, f_{\alpha\beta}, \ \alpha,\beta \in I, \ (I,\leq))$$

and define for $x, y \in \Pi X_\alpha$

$$x \approx y :\Longleftrightarrow x_\alpha = y_\alpha \text{ for all standard } \alpha.$$

The classical description of the convergence in this uniformity is given by considering (Moore - Smith -) sequences which coincide on larger and larger initial segments of $(I,\leq)$.
If in addition all X_α are finite, then the standard X_α contain only standard elements. If we now take some $\gamma \geq \alpha$ for all standard

$\alpha \in I$, then the projections on all standard X_α are standard and using the axiom (S) we obtain a standard element of the inverse limit $X = \varprojlim X_\alpha$; in particular X is non - empty. If all f_α are surjectiv we obtain:

Proposition 1:

$X = \varprojlim X_\alpha$ is the (continuous, homomorphic etc.) image of each X_γ for which $\alpha \leq \gamma$ for all standard α.

Hence in some sense the profinite structures can be replaced by non-standard finite structures. This has been originated by A. Robinson; cf. also work of Hirshfeld and Manevitz for "lifting" theorems about finite groups to profinite groups (infinite Galois theory, prosolvable groups, pro - nilpotent groups etc.).

§4 Direct limits:

In the same way as inverse limits are replaced by structures with an infinite index this should be done for direct limits. Take for simplicity a strictly increasing standard sequence

$(X_n, n \in \mathbb{N}) \quad X_n \subsetneqq X_m, n < m.$

We could think e. g. of topological vector spaces and continuous inclusions; this means that the monads μ_n form a strictly increasing sequence too. Generally one has the feeling that direct systems are "nicer" than inverse systems; unfortunately the opposite is true for the case of monads:

The intersection of monads is again a monad by the theorem on monads but the union of monads is in general not a monad. How to describe the smallest monad containing $\bigcup_{n \in \mathbb{N}} \mu_n$?

Given a standard family $(\mathscr{F}_i \mid i \in I)$ of filters on sets X_i, the product filter $\mathscr{F} = \Pi(\mathscr{F}_i \mid i \in I)$ projects on each set X_i ($i \in I$ standard or nonstandard); the same is true for the monad $\mu = \mu_{\mathscr{F}}$.

There the projections $P_i: \prod_{i \in I} X_i \to X_i$ yield:

$$P_i(\mu) = \mu_i = \mu_{\mathscr{F}_i} \quad \text{for standard } i \in I;$$

for nonstandard $i \in I$ we put

$$^{\Pi}\mu_i = {}^{\Pi}\mu_i(\mathscr{F}) = P_i(\mu)$$

and call it the Π - monad at $i \in I$ w. r. t. $(\mathcal{F}_i \mid i \in I)$ (B. Benninghofen); for standard $i \in I$ we use ${}^{\Pi}\mu_i$ instead of μ_i too.

This allows us to describe the supremum of monads:

Proposition 2: $\mu = \bigcup({}^{\Pi}\mu_i \mid i \in I)$ is the smallest monad containing all monads μ_i, $i \in I$, i standard.

Intuitively this means that some "very small monads", namely the ${}^{\Pi}\mu_i$ for nonstandard $i \in I$ are added. This technique is particularly useful in the case of strict direct limits of locally convex topological vector spaces $(V_n, n \in \mathbb{N})$, e. g. when one wants to prove the completeness of the limit space if all V_i are complete.
Another application is the explicit construction of the truth value object and the exponent objects in the topos of sheaves over some topological space where classically the direct limit constructions are hidden in the applications of the associated sheave functor.

Literature:

B. Benninghofen: Infinitesimals and Superinfinitesimals. Dissertation Aachen 1982

B. Benninghofen - M. M. Richter: General theory of Superinfinitesimals.Submitted for publication.

Y. Hirshfeld - L. M. Manevitz: Profinite and * - finite groups Preprint 1981.

W. A. J. Luxemburg: A General Theory of Monads. In: Applications of Model Theory to Algebra, Analysis and probability, 1964 (Ed. W. A. J. Luxemburg), p. 18 - 86.

E. Nelson: Internal set theory. Bull AMS 83 (1977), p. 1165 -1198.

M. M. Richter: Ideale Punkte, Monaden und Nichtstandard - Methoden, Wiesbaden 1982.

M. M. Richter - M. E. Szabo: Towards a Nonstandard Analysis of Programs.In: Nonstandard Analysis - Recent Developments, ed. A.E. Hurd, SLN 983,p.186-2o3.

A. Robinson: Non - standard Analysis. Amsterdam 1966.

A. Robinson: Non - standard Arithmetic.
Bull. AMS 73 (1967), p. 818 - 843.

D. Scott: Continuous lattices.
In: Toposes, Algebraic Geometry and Logic, 1971 (ed. F. W. Lawvere), p. 97 - 136.

Universal Algebra and its Links with Logic,
Algebra, Combinatorics, and Computer Science
Proc. "25. Arbeitstagung über Allgemeine Algebra", Darmstadt 1983
P. Burmeister et al. (eds.)

85 – 105

ASPECTS OF QUANTIFIER ELIMINATION IN ALGEBRA

V. Weispfenning

Quantifier elimination (q.e.) is one of the oldest and most powerful techniques in mathematical logic and above all in algebraic model theory. It is relevant to all of the following model-theoretic and algebraic topics: Classification of structures according to elementary equivalence (elementary invariants), transfer of elementary properties (persistence, model-completeness, substructure-completeness), characterization of definable sets and relations in structures, stability, decidability, $\aleph_0$-categoricity, amalgamation, extension of morphisms, homogeneous and universal structures, algebraic and existential closedness, Nullstellensätze, definite functions and the like.

The goal of this note is to give an introduction and survey of the method and its algebraic significance in general. It is written for the algebraically minded reader acquainted with the basic concepts of model theory. To encourage the potential user, the presentation of the fundamental techniques is nearly self-contained. §1 introduces the basic syntactical notions with some motivation and historical notes. §2 presents the semantical criteria for q.e. in terms of persistence and model-completeness, originating with A.Robinson. §3 continues with criteria in terms of extensions of morphisms and saturated structures. §4 reviews some characterizations of q.e. structures for specific algebraic theories. Due to limitations of space, we will not discuss the considerable body of results concerning q.e. in infinitary languages (as related e.g. to existentially closed and generic structures) and languages with generalized quantifiers. Results on q.e. for specific algebraic theories will be mentioned as we go along.

§ 1 SYNTACTICAL REDUCTION AND THE PRE-1950 PERIOD.

Consider the following example from field theory: Let $\underline{F}$ be a field, $f(X), g(X)$ monic polynomials in $\underline{F}[X]$ of fixed degrees m, n, respectively. Then f and g have a common zero in a suitable extension field $\underline{F}'$ of $\underline{F}$ iff the resultant $R(f,g)$ of f and g vanishes. Equivalently, for an algebraically closed field $\underline{F}$,

$$\underline{F} \models \exists x(f(x)=0 \wedge g(x)=0) \leftrightarrow R(f,g)=0 .$$

Notice that R(f,g) is a polynomial over $\mathbb{Z}$ in the coefficients of f and g, formed uniformly and effectively without reference to the field F; in particular, these coefficients may be taken as indeterminates. In other words, the assignment $\exists x(f(x)=0 \wedge g(x)=0) \longmapsto R(f,g)=0$ reduces a (semantically) "complicated" formula to a "simple" formula in the class of all algebraically closed fields. The claim the "R(f,g)=0" - and in fact any quantifier-free (q.f.) formula φ - is "simple" can be substantiated in two complementary ways:

(i) Testing the validity of φ involves only the most basic features of the theory: execution of the ring operations, testing on equality, calculation of truth--values for the propositional connectives $\wedge$ (and), $\vee$ (or), $\neg$ (not), $\rightarrow$ (implies), $\leftrightarrow$ (equivalent).

(ii) φ is preserved under any morphism $h: F' \rightarrow K$ from a subfield of F containing all parameters of φ (coefficients of f and g) into any other field K.

This example is a simple instance of the pre-1930 "algebraic" elimination theory (cf. Wb 1898 , Wa 1931). Phrased in terms of first-order logic, this elimination theory deals with effective and uniform reductions of existential formulas to quantifier-free (possibly infinitary) formulas in algebraically closed fields and related situations. (If L is a first-order language, then an atomic L-formula is an equation $t_1 = t_2$ or a relation $R(t_1,\ldots,t_n)$, where t_i are L-terms and R is a relation-symbol in L. The closure of the class At of atomic L-formulas under $\neg$; $\wedge, \vee$; $\neg, \wedge, \vee$ is the class Ba of basic L-formulas, the class Q_0^+ of positive q.f. L-formulas, the class Q_0 of q.f. L-formulas, respectively. The closure of Q_0 (Q_0^+) under existential quantifiers $\exists x$ is the class E (E^+) of (positive) existential L-formulas. The closure of Q_0 under universal quantifiers $\forall x$ is the class U of universal L-formulas.)

A prominent example in algebraic elimination theory is (the effective version of) Hilbert's Nullstellensatz which provides a q.f. equivalent for formulas of the form $\exists \underline{x}(f_1(\underline{x})=0 \wedge \ldots \wedge f_m(\underline{x})=0 \wedge g(\underline{x}) \neq 0)$, where $\underline{x}$ denotes a finite tuple of variables. A counterpart in the theory of real-closed fields is Sturm's theorem dealing with the number of roots of a polynomial in an interval.

Once that the concept of an arbitrary first-order formula is at hand, a simple observation shows that an elimination theory for existential formulas immediately yields an elimination theory for all formulas. We place this observation in a slightly more general context: Let Φ, Ψ denote sets of L-formulas. We shall tacitly assume that all the sets considered contain all equations "x=y" between variables and are closed under substitution of variables.

Then $\neg\Psi, \exists\Psi, \forall\Psi$ is the set of all formulas of the form $\neg\psi, \exists x\psi, \forall x\psi$, respectively, with $\psi \in \Psi$. $\bigwedge\Psi, \bigvee\Psi, \mathcal{B}\Psi, \underline{\exists}\Psi, \underline{\forall}\Psi, \mathcal{F}\Psi$ is the closure of Ψ under conjunction, disjunction, propositional connectives, existential quantifiers, universal quantifiers, propositional connectives and quantification, respectively. $Fo = \mathcal{F}At$ is the set of all L-formulas. Let K be a class of L-structures, $\varphi \in Fo$, $\Phi, \Psi \in Fo$. Then we say: φ is reducible to Ψ in K, if there exists $\varphi' \in \Psi$ such that $K \models \varphi \leftrightarrow \varphi'$. Φ is (recursively, primitive recursively) reducible to Ψ in K, if every $\varphi \in \Phi$ is reducible to Ψ in K (and Φ and the reduction $\varphi \mapsto \varphi'$ is (primitive) recursive). K admits (recursive, prim. recursive) quantifier elimination relative to Ψ, if Fo is (recursively, prim. recursively) reducible to Ψ in K. If $\Psi = Q_o$, the reference to Ψ is omitted.

PROPOSITION 1.1 (i) $\underline{\exists}\Psi$ ($\underline{\forall}\Psi$) is (rec.ly, pr. rec.ly) reducible to Ψ in K iff $\exists\Psi$ ($\forall\Psi$) is (rec.ly, pr. rec.ly) reducible to Ψ in K.
(ii) Let $At \cup \neg\Psi \subseteq \Psi$. Then K admits (rec., pr. rec.) q.e. relative to $\bigwedge\bigvee\Psi$ iff $\exists\bigwedge\Psi$ is (rec.ly, pr. rec.ly) reducible to $\bigwedge\bigvee\Psi$ in K.
(iii) Let $At \cup \neg\Psi \subseteq \Psi$. Then K admits (rec., pr. rec.) q.e. relative to $\underline{\exists}\bigwedge\bigvee\Psi$ iff $\underline{\forall}\bigvee\Psi$ is (rec.ly, pr. rec.ly) reducible to $\underline{\exists}\bigwedge\bigvee\Psi$ in K.

PROOF. Induction on the number of quantifiers ((i),(ii)), and on the number of quantifier-blocks $\underline{\exists}\underline{x}$, $\underline{\forall}\underline{x}$ in prenex formulas.

COROLLARY 1.2 (i) E^+ is (rec.ly, pr. rec.ly) reducible to Q_o^+ iff $\exists\bigwedge At$ is (rec.ly, pr. rec.ly) reducible to Q_o^+.
(ii) K admits (rec., pr. rec.) q.e. iff $\exists\bigwedge Ba$ is (rec.ly, pr. rec.ly) reducible to Q_o in K.
(iii) K admits (rec., pr. rec.) q.e. relative to E iff $\underline{\forall}\bigvee Ba$ is (rec.ly, pr. rec.ly) reducible to E in K.

The pre-195o period in quantifier elimination is characterized by a number of explicit, prim. recursive reduction procedures for specific algebraic theories, and their application to decidability and elementary invariants (cf. 1.3, 1.4 below). The procedures combine 1.2(ii) with an algebraic elimination theory for $\exists\bigwedge$ Ba-formulas designed specifically for (or adapted from) the algebraic theory in question. It should be noted that many of these q.e. procedures take place in an extension L' (by definable relations and functions) of the natural algebraic language L of the theory. So, regarded from within L, these are q.e. procedures relative to certain sets Ψ of L-formulas that may be considerably larger than Q_o.

The following algebraic classes K were treated in this period:

Atomic boolean algebras (Sko 1919), 'Morleyzation' (Sko 1920), dense and discrete linear orders (La 1927, Sko 1928), Skolem functions (Sko 1928), additive arithmetic (Pb 1929), additive arithmetic and Z-groups (Sko 193o), multiplicative arithmetic of number fields (q.e. for sentences, Sko 1930), boolean algebras (Ta 1949), algebraically and real closed fields (Ta 1949), well-orders (MoTa 1949), abelian groups (Sm 1949).

For later papers in the same spirit see e.g.: Bö 1983, BoCh 1980, Br 1978, Co 1969, Cl 1983, Cm 1976, Cp 1972, FrRk 1975, FeVa 1959, FrSc 1976, Gu 1977, KrKr 1967, KvKn 1969, Mk 1975,, Mv 1971, Po 1983, Ra ,Scht, Se 1954/56a, Sm 1955, TaMK 1951, Vo 1976, We 1971/75/75a/76/81/82/b/c .

We indicate some immediate syntactical consequences of a q.e. procedure.

PROPOSITION 1.3 Suppose K admits q.e. relative to Ψ and $\underline{A} \in K$. Then:
(i) Any $S \subseteq A^n$ definable (with parameters) in $\underline{A}$ is definable (with parameters) in $\underline{A}$ be a formula in Ψ.
(ii) K is complete iff K is Ψ-complete.

(K is (Ψ-)complete if for all sentences ψ ($\psi \in \Psi$), $K \models \psi$ or $K \models \neg\psi$.)

PROPOSITION 1.4 Suppose K admits rec. (pr. rec.) q.e. relative to $\wedge\vee\Psi$ [relative to $\exists\wedge\vee\Psi$, and $\neg\Psi \subseteq \Psi$] . Then:
(i) K is (pr. rec.ly) decidable iff K is (pr. rec.ly $\vee\Psi$-decidable [(pr. rec.ly) $\underline{\forall}\vee\Psi$-decidable] .
(ii) K is (pr. rec.ly) decidable and complete iff K is (pr. rec.ly) Ψ-decidable and Ψ-complete [K is (pr. rec.ly) $\underline{\exists}\wedge\Psi$-decidable and $\underline{\exists}\wedge\Psi$-complete] .

(K is (pr. rec.ly) [Ψ-]decidable, if $\{\psi$: ψ sentence [in Ψ], $K \models \psi\}$ is (prim.) recursive.) For incomplete q.e. classes K, 1.3(ii) can be used to find all complete subclasses of K by reading off elementary invariants from the atomic L-sentences. 1.3 and 1.4 clarify the special importance of the choice of language in absolute q.e. - or equivalently of the choice of Ψ in relative q.e. - for applications. In fact, experience shows that the selection of an appropriate set Ψ is one of the most delicate problems in q.e., e.g. for p-adic fields, modules and abelian groups (cf. Ma 1976 , We 1976 , Ba 1976 , Mk , Sm 1955 , EkFi 1972). To reinforce this point, we sketch two ways to obtain q.e. for any class K of L-structures by "brute force" (cf. Sko 1920, Sko 1928, MrVa 1962).

PROPOSITION 1.5 (i) (Morleyzation) Obtain L' from L by adding a new relation-symbol R_φ for any $\varphi(x,\underline{y}) \in Q_0(L)$ with defining axiom $\exists x\, \varphi(x,\underline{y}) \leftrightarrow R_\varphi(\underline{y})$.
(ii) (Skolemization) Obtain L' from L by adding a new function-symbol f_φ for any $\varphi(x,\underline{y}) \in Q_0(L)$ with defining axiom $\exists x\, \varphi(x,\underline{y}) \leftrightarrow \varphi(f_\varphi(\underline{y}),\underline{y})$.
(iii) In both (i) and (ii) put $L^* = \bigcup L_n$, where $L_0 = L$, $L_{n+1} = L_n'$.
Then K admits q.e. in L^* .

While these procedures are convenient tools in general model theory, they are virtually useless to obtain information about elementary properties of algebraic theories.

§ 2 PERSISTENCE AND MODEL-COMPLETENESS.

A new line of ideas was introduced into q.e. by A.Robinson in the 195o's. He used non-constructive methods - the compactness theorem for first-order logic and diagrams of structures- to derive indirect criteria for the existence of (relative) q.e. in terms of persistence of formulas under certain morphisms. This approach has had a profound influence on algebraic model theory and has stimulated many deep theorems (cf. e.g. AdKi 1976, AxKo 1966, Bs 1979, Ba 1976, De , Dr 1978, Er 1967/80, Ja 1980, JdKi 1975, Pl 1981/82, Schd 1982,Wh 1979a, Zi 1972).

In Robinson's approach, the whole body of structural algebraic results of the theory can be applied directly to prove persistence theorems; in the syntactic approach the available algebraic information has essentially to be reproved with special attention on uniformity and effectiveness. So the dichotomy between "old" (syntactical) algebra and "modern" (structural) algebra has a counterpart in these two viewpoints on quantifier elimination. There is of course an a priori loss of effectivity in the structural approach; frequently, however, this can be repaired by an application of Gödel's completeness theorem - at least, when "effective" is equated with "general recursive":

PROPOSITION 2.1 Let K be a class of L-structures axiomatized by a recursively enumerable axiom-system Θ , and let Φ , Ψ be sets of L-formulas.
(i) If K is Φ-complete and Φ is recursive, then K is Φ-decidable.
(ii) If Φ is reducible to Ψ in K and Φ, Ψ are recursive, then Φ is recurively reducible to Ψ in K.
(iii) If K admits q.e. relative to Ψ and Ψ is recursive, then K admits recursive q.e. relative to Ψ .

PROOF. (i) Let φ be a sentence in Φ; enumerate all proofs from Θ until a proof for φ or a proof for $\neg\varphi$ is reached. (ii) Let $\varphi \in \Phi$. Enumerate all proofs from Θ until a proof for a formula of the form $\varphi \leftrightarrow \psi$ with $\psi \in \Psi$ is reached. (iii) Apply (ii) to $\Phi = \mathrm{Fo}$.

On the other hand, for more specific questions on effectivity an explicit prim. recursive q.e. procedure is a much richer source of information than a proof by compactness (e.g. for effective Nullstellensätze, cf. Se 1956, a).

Since the 196o's, Robinson's method was extended and refined by many authors : cf. e.g. Bu 1975, Dr 1978/82, EkSb 1971, HiWh 1975, Ho 1980, Ke 1960, Ko 1962, Kr 1976/77, Li 1964, Me 1979, Ro 1963, Sa 1972, Sr 1973, Sh 1967/71/77, We 1978, Wh 1978, Wi 1975. Besides, there is a large part of general model theory dealing with syntactical reduction of formulas preserved under various algebraic constructions that will not be touched upon here, cf. [ChKe 1973] . In the rest of this section and in section 3 we present the semantical criteria for (relative) q.e. in a condensed version.

All classes K of L-structures we consider are assumed to be closed under isomorphisms. Call K ∞-compact (compact), if for every extension L' of L by (finitely many) new constants and every set Φ of sentences in L', if every finite subset Ψ of Φ has a model $\underline{A}_\Psi$ with $\underline{A}_\Psi | L \in K$, then Φ has a model $\underline{A}$ with $\underline{A} | L \in K$. By the compactness theorem for first-order logic every elementary class K (i.e. every class K axiomatized by a set of L-sentences) is ∞-compact. If $\underline{A}, \underline{B} \in K$, A' is a subset of the universe A of $\underline{A}$, then a morphism $\underline{A} \supseteq A' \xrightarrow{f} \underline{B}$ is a map $f: A' \to B$; if A' = A, we write $\underline{A} \xrightarrow{f} \underline{B}$. If $\varphi(\underline{x}) \in \mathrm{Fo}$, we say "f preserves φ" or "φ is persistent under f ", if for all tuples $\underline{a}$ in A', $\underline{A} \models \varphi(\underline{a})$ implies $\underline{B} \models \varphi(f\underline{a})$. More generally, if $\varphi(\underline{x}), \varphi'(\underline{x}) \in \mathrm{Fo}$, we say " f preserves (φ, φ')" or (φ,φ') is persistent under f ", if for all $\underline{a}$ in A', $\underline{A} \models \varphi(\underline{a})$ implies $\underline{B} \models \varphi'(f\underline{a})$. So f preserves φ iff it preserves (φ, φ). $\Phi \subseteq \mathrm{Fo}$ is preserved under f if every $\varphi \in \Phi$ is preserved under f . So $\underline{A} \xrightarrow{f} \underline{B}$ is a homomorphism (an embedding) iff it preserves At (Ba). $\underline{A} \subseteq \underline{B}$ is a Φ-extension, $\underline{A} \overset{\Phi}{\leq} \underline{B}$, if $\mathrm{id}: \underline{A} \to \underline{B}$ preserves Φ . If $A' \subseteq A \cap B$, then $\underline{A}$ and $\underline{B}$ are Φ-equivalent over A', $\underline{A} \overset{\Phi}{\underset{A'}{\equiv}} \underline{B}$, if $\underline{A} \supseteq A' \xrightarrow{\mathrm{id}} \underline{B}$ and $\underline{B} \supseteq A' \xrightarrow{\mathrm{id}} \underline{A}$ are Φ-preserving. If in these concepts $\Phi = \mathrm{Fo}$, it is omitted in the notation (elementary extension, elementary equivalence over A'); if $A' = \emptyset$ it is omitted too.

We formulate two fundamental persistence theorems as interpolation-style theorems.

FIRST PERSISTENCE THEOREM 2.2 Let $\varphi, \varphi' \in Fo, \Psi \subseteq Fo$, K a ∞-compact class of L-structures, and suppose there is a sentence $\sigma \in \exists\bigwedge\Psi$ with $K \models \neg\sigma$. Then (φ, φ') is persistent under all Ψ-preserving morphisms $\underline{A} \xrightarrow{f} \underline{B}$ with $\underline{A}, \underline{B} \in K$ iff there exists $\psi \in \exists\bigwedge\bigvee\Psi$ such that $K \models (\varphi \to \psi) \wedge (\psi \to \varphi')$. In particular, φ is persistent under all Ψ-preserving morphisms $\underline{A} \xrightarrow{f} \underline{B}$ with $\underline{A}, \underline{B} \in K$ iff φ is reducible in K to $\exists\bigwedge\bigvee\Psi$.

SECOND PERSISTENCE THEOREM 2.3 Let $\varphi, \varphi' \in Fo$, $\Psi \subseteq Fo$, K a compact class of L-structures, and suppose there are sentences $\tau, \sigma \in \bigwedge\bigvee\Psi$ with $K \models \tau \wedge \neg\sigma$. Then (φ, φ') is persistent under all Ψ-preserving morphisms $\underline{A} \supseteq A' \xrightarrow{f} \underline{B}$ with $\underline{A}, \underline{B} \in K$ iff there exists $\psi \in \bigwedge\bigvee\Psi$ such that $K \models (\varphi \to \psi) \wedge (\psi \to \varphi')$. In particular, φ is persistent under all Ψ-preserving morphisms $\underline{A} \supseteq A' \xrightarrow{f} \underline{B}$ with $\underline{A}, \underline{B} \in K$ iff φ is reducible to $\bigwedge\bigvee\Psi$ in K.

Since the proofs are similar, we give only the nontrivial direction of 2.2 : Suppose $(\varphi(\underline{x}), \varphi'(\underline{x}))$ is persistent under all Ψ-preserving $\underline{A} \xrightarrow{f} \underline{B}$ with $\underline{A}, \underline{B} \in K$, and put $\Phi = \{\psi \in \exists\bigwedge\Psi : K \models \psi \to \varphi'\} \neq \emptyset$. Then $K \models \bigvee\Phi \to \varphi'$, and so by compactness it suffice to show: $(*)$ $K \models \varphi \to \bigvee\Phi$.
Assume $(*)$ fails; then there exists $\underline{A} \in K$, $\underline{a}$ in A with $\underline{A} \models \varphi(\underline{a}) \wedge \neg\bigvee\Phi(\underline{a})$, So for any Ψ-preserving morphism $\underline{A} \xrightarrow{f} \underline{B} \in K$, $\underline{B} \models \varphi'(f\underline{a})$. For $A' \subseteq A$, define the diagram $Diag_{\Psi}(\underline{A}, A')$ as the set of all sentences $\psi[\dot{\underline{a}}]$, obtained from $\psi(\underline{x}) \in \Psi$ by substitution of names for elements of A', which are true in $\underline{A}$. Then $Diag_{\Psi}(\underline{A}, A) \cup \{\neg\varphi'[\dot{\underline{a}}]\}$ has no model $\underline{B}$ with $\underline{B} \mid L \in K$. By ∞-compactness there exists $\rho(\underline{x}, \underline{y}) \in \bigwedge\Psi$ and $\underline{b}$ in A such that $\{\rho[\dot{\underline{a}}, \dot{\underline{b}}], \varphi'[\dot{\underline{a}}]\}$ has no such model. We may assume that $\underline{a} = (a_1, \dots, a_n)$, that $a_1, \dots, a_r$ are different in A and that $\{a_1, \dots, a_n\} = \{a_1, \dots, a_r\}$ for some $1 \le r \le n$. Let $\delta(x_1, \dots, x_n) \in \bigwedge\Psi$ say that $\{x_1, \dots, x_n\} = \{x_1, \dots, x_r\}$. Then $K \models \forall\underline{x}\forall\underline{y}(\delta \wedge \rho \to \varphi')$, and so $K \models \forall\underline{x}(\exists\underline{y}(\delta \wedge \rho) \to \varphi')$, and so $\exists\underline{y}(\delta \wedge \rho) \in \Phi$, and so $\underline{A} \models \neg\exists\underline{y}(\delta \wedge \rho)(\underline{a})$. On the other hand, by definition of ρ and δ, $\underline{A} \models \exists\underline{y}(\delta \wedge \rho)(\underline{a})$, a contradiction.

Note for the model theorist: 2.2 and 2.3 are indeed related to classical interpolation. E.g. Craig's interpolation lemma and Beth's definability lemma can be derived from 2.3 via the Keisler-Shelah theorem (ChKe [1973] , 6.1.15).

Taking $\Psi = Q_o, Q_o^+$ we obtain the following corollaries.

COROLLARY 2.4 (i) Let K be ∞-compact (and assume $K \models \neg\sigma$ for some positive existential sentence σ). Then φ is equivalent in K to a (positive) existential formula iff φ is persistent under all embeddings (homomorphisms) between structures in K.

(ii) Let K be compact and suppose there is a constant 0 in L (and a constant 1 in L with $K \models 0 \neq 1$). Then φ is equivalent in K to a formula in Q_0 (Q_0^+) iff φ is persistent under all morphisms $\underline{A} \supseteq A' \xrightarrow{f} \underline{B}$, $\underline{A},\underline{B} \in K$, preserving Ba (At).

To formulate the global versions of 2.2 and 2.3 we define <u>relative model-completeness and substructure completeness</u>. K is <u>Ψ-model-complete</u> if for all $\underline{A} \overset{\Psi}{\lesssim} \underline{B}$ in K, $\underline{A} \preceq \underline{B}$. K is <u>Ψ-substructure-complete</u> if for all $\underline{A},\underline{B} \in K$, $A' \subseteq A \cap B$, if $\underline{A} \overset{\Psi}{\equiv}_{A'} \underline{B}$ then $\underline{A} \equiv_{A'} \underline{B}$. For Ψ = Ba, the reference to Ψ is omitted. Then 2.2 and 2.3 together with 1.1 yield 2.5 and 2.6.

<u>THEOREM</u> 2.5 (<u>Robinson's test</u>) Let K be ∞-compact, $At \cup \neg\Psi \subseteq \Psi$. Then t.f.a.e.:

(i) K is Ψ-model-complete .

(ii) K admits q.e. relative to $\exists\wedge\vee\Psi$.

(iii) Every Ψ-extension $\underline{A} \overset{\Psi}{\lesssim} \underline{B}$ in K preserves $\forall\vee\Psi$.

<u>THEOREM</u> 2.6 Let K be compact, $At \cup \neg\Psi \subseteq \Psi$,and assume L contains a constant. Then the following are equivalent:

(i) K is Ψ-substructure-complete.

(ii) K admits q.e. relative to $\wedge\vee\Psi$.

(iii) For all $\underline{A},\underline{B} \in K$, $A' \subseteq A \cap B$, if $\underline{A} \overset{\Psi}{\equiv}_{A'} \underline{B}$ then $\underline{A} \overset{\exists\wedge\Psi}{\equiv}_{A'} \underline{B}$ (It suffices in fact to consider finite sets A').

<u>COROLLARY</u> 2.7 Let $At \cup \neg\Psi \subseteq \Psi$, and let K be a Ψ-model-complete, elementary class. Then K can be axiomatized by axioms in $\forall\exists\wedge\vee\Psi$.

<u>PROOF</u>. Let Φ_1 be the set of all sentences in $\exists\wedge\vee\Psi$ that are true in K, and let Φ_2 be the set of all equivalences $\varphi \leftrightarrow \psi$ with $\varphi \in \forall\vee\Psi$, $\psi \in \exists\wedge\vee\Psi$, that are true in K. Then Φ_2 is logically equivalent to a set Φ_3 of sentences in $\forall\exists\wedge\vee\Psi$. Let $K' = \mathrm{Mod}(\Phi_1 \cup \Phi_3)$. Then by 2.5, $K' \supseteq K$, and by 1.1(iii), K' admits q.e. relative to $\exists\wedge\vee\Psi$. Consequently, K' = K .

The following partial converse of 2.7 is due to Lindström for Ψ = Ba .

<u>THEOREM</u> 2.8 Let K be axiomatized by a set of L-sentences in $\forall\exists\wedge\vee\Psi$, let $At \cup \neg\Psi \subseteq \Psi$, κ an infinite cardinal $\geq$ card(L), and assume K contains no finite structure. If any two structures in K of cardinality κ are isomorphic, then K is Ψ-model-complete.

PROOF. By the theorem of Löwenheim-Skolem-Tarski, it suffices to prove condition 2.5(iii) for structures $\underline{A}$, $\underline{B}$ of cardinality κ. Call $\underline{C} \in K$ $\exists\wedge\Psi$-closed in K, if $\underline{C} \overset{\Psi}{\leq} \underline{D} \in K$ implies $\underline{C} \overset{\forall\vee\Psi}{\leq} \underline{D}$. By forming a union of a suitable chain of structures in K, we find $\underline{A} \overset{\Psi}{\leq} \underline{C} \in K$ such that $\underline{C}$ is $\exists\wedge\Psi$-closed in K and has cardinality κ. Since $\underline{A} \cong \underline{C}$, $\underline{A}$ is $\exists\wedge\Psi$-closed as well, which proves condition 2.5(iii).

We close this section with some easy facts on q.e., completeness and amalgamation. We say K admits Ψ-amalgamation if for all $\underline{A} \overset{\Psi}{\underset{A'}{\equiv}} \underline{B}$, $\underline{A},\underline{B} \in K$ there exist $\underline{C} \in K$ and Ψ-preserving morphisms $\underline{A} \xrightarrow{f} \underline{C}$, $\underline{B} \xrightarrow{f} \underline{C}$ with $f \mid A' = g \mid A'$. So K admits Ba-amalgamation iff the class Sub(K) of all substructures of structures in K has the amalgamation property in the usual sense.

PROPOSITION 2.9 Let K be ∞-compact, $At \cup \neg\Psi \subseteq \Psi$. Then K is Ψ-substructure-complete iff K is Ψ-model-complete and admits Ψ-amalgamation.

PROOF. If Ψ = Ba one obtains the usual notion of model-completeness and substructure-completeness. The proof can now be adapted from EkSb [1971].

Call $\underline{A} \in K$ Ψ-prime if for every $\underline{B} \in K$ there exist a Ψ-preserving morphism $\underline{A} \xrightarrow{f} \underline{B}$.

PROPOSITION 2.10 (Prime model test) Let $At \cup \neg\Psi \subseteq \Psi$. If K is Ψ-model-complete and has a Ψ-prime structure, then K is complete.

A similar argument together with 2.6 yields:

PROPOSITION 2.11 (Prime extension test) Let $At \cup \neg\Psi \subseteq \Psi$, K compact. If every Ψ-extension in K is a $\exists\wedge\Psi$-extension, and for all $\underline{A},\underline{B} \in K$, all (finite) $A' \subseteq A \cap B$ with $\underline{A} \overset{\Psi}{\underset{A'}{\equiv}} \underline{B}$ there exists $\underline{A} \overset{\Psi}{\leq} \underline{C} \in K$, $C \supseteq A'$, and a Ψ-preserving $\underline{C} \xrightarrow{f} \underline{B}$ with $f \mid A' = id$, then K is Ψ-substructure-complete, and so admits q.e. relative to $\wedge\vee\Psi$.

REMARK 2.12 K is Ψ-substructure-complete iff every complete subclass of K is Ψ-substructure-complete and for all $\underline{A},\underline{B} \in K$, $A' \subseteq A \cap B$, $\underline{A} \overset{\Psi}{\underset{A'}{\equiv}} \underline{B}$ implies $\underline{A} \equiv \underline{B}$.

For certain classes K, model-completeness implies substructure-completeness:

PROPOSITION 2.13 [BdLn 1973]. Let K be an elementary class closed under the formation of substructures and direct products (i.e. a universal Horn class) and let K' be the class of all infinite structures in K. If K' is model-complete, then K' is substructure-complete.

PROOF. Let $\underline{A}, \underline{B} \in K'$, $\underline{A}' \subseteq \underline{A}, \underline{B}$, $\underline{A} \overset{Ba}{\underset{A'}{\equiv}} \underline{B}$, let $f: \underline{A} \to \underline{A} \times \underline{A}$, $g: \underline{B} \to \underline{B} \times \underline{B}$ be the diagonal embeddings and let $A'' = f(A') = g(A')$. Then $\underline{A} \times \underline{A} \succeq \underline{A} \times \underline{A}' \preceq \underline{A} \times \underline{B}$, $\underline{A} \times \underline{B} \succeq \underline{A}' \times \underline{B} \preceq \underline{B} \times \underline{B}$, and so $\underline{A} \times \underline{A} \equiv_{A''} \underline{B} \times \underline{B}$. Since f and g are elementary embeddings, this shows that $\underline{A} \equiv_{A'} \underline{B}$.

§ 3 EXTENSION OF MORPHISMS.

We continue the semantical characterizations of q.e. by some criteria related to existence and extensions of morphisms and saturated structures. Call $\underline{A}$ homogeneous if every isomorphism between finitely generated (f.g.) substructures of $\underline{A}$ extends to an automorphism of $\underline{A}$. $\underline{A}$ is ω-homogeneous if for all f.g. substructures $\underline{B} \subseteq \underline{B}'$ of $\underline{A}$ any embedding $\underline{B} \xrightarrow{f} \underline{A}$ extends to an embedding $\underline{B}' \xrightarrow{f} \underline{A}$. $\underline{A}$ is locally finite if every f.g. substructure of $\underline{A}$ is finite. $\underline{A}$ is ω-universal in K if $\underline{A} \in K$ and if every f.g. substructure $\underline{B}$ of some $\underline{C} \in K$ can be embedded into $\underline{A}$. Notice that a denumerable $\underline{A}$ is homogeneous iff it is ω-homogeneous. (Extend embeddings stepwise, going back and forth.)

PROPOSITION 3.1 Let L be finite, K an At-complete, elementary class consisting of locally finite L-structures. Then K admits q.e. iff every $\underline{A} \in K$ is ω-homogeneous and ω-universal in K.

PROOF. $\Rightarrow$: Let $\underline{B} = \langle \underline{b} \rangle \subseteq \underline{C} \in K$ and let $\varphi(\underline{x}) \in \wedge Ba$ describe the isomorphism-type of $\underline{B}$, $\underline{C} \models \varphi(\underline{b})$. Then $\underline{C} \models \exists \underline{x} \varphi$ and so $\underline{A} \models \exists \underline{x} \varphi$, and so $\underline{C}$ embeds into $\underline{A}$. Next, let $\langle \underline{b} \rangle \cong \langle \underline{c} \rangle$ in $\underline{A}$ via f with $f(\underline{b}) = \underline{c}$, and let $a \in A$. Then there is a $\psi(x,\underline{y}) \in \wedge Ba$ describing the isomorphism-type of $\langle a,\underline{b} \rangle$, $\underline{A} \models \psi(a,\underline{b})$. Then $\underline{A} \models \exists x \psi(x,\underline{b})$ and so $\underline{A} \models \exists x \psi(x,\underline{c})$, say $\underline{A} \models \psi(a',\underline{c})$. Extend f, mapping a onto a'.
$\Leftarrow$: Let $\underline{A},\underline{B} \in K$, $\langle \underline{c} \rangle \subseteq \underline{A},\underline{B}$, $\varphi(x,\underline{y}) \in \wedge Ba$ such that $\underline{A} \models \exists x \varphi(x,\underline{c})$, say $\underline{A} \models \varphi(a,\underline{c})$. Then by assumption there is an embedding $\langle a,\underline{c} \rangle \xrightarrow{f} \underline{B}$ with $f | \langle \underline{c} \rangle = id$, and so $\underline{B} \models \exists x \varphi(x,\underline{c})$. By 2.6, this shows that K admits q.e..

An L-structure $\underline{A}$ is uniformly locally finite if for some function $f: \omega \to \omega$, any n-generated substructure of $\underline{A}$ has at most $f(n)$ elements. In particular any finite $\underline{A}$ is uniformly locally finite.

COROLLARY 3.2 Let L be finite and let $\underline{A}$ be a (uniformly locally) finite L-structure. Then $\underline{A}$ admits q.e. iff $\underline{A}$ is (ω-) homogeneous.

PROOF. Apply 3.1 to the class $K = \{\underline{B} : \underline{B} \equiv \underline{A}\}$.

REMARK 3.3 (i) If the equivalent conditions of 3.1 are satisfied, then a back--and-forth extension of embeddings shows that any two countable infinite $\underline{A},\underline{B} \in K$ are isomorphic, i.e. K is $\aleph_0$-categorical .
(ii) It is not possible to replace ω-universality in 3.1 by the amalgamation property for Sub(K) (even if all $\underline{A} \in K$ are infinite) as claimed in [Kr 1975] , Cor. 16 . Counterexample: Let K be the class of all infinite sets A with a distinguished subset U containing at most one element.

Without local finiteness a counterpart to 3.1 requires suitable elementary extensions or structures of enough saturation (see [Sa 1972] for the definition of κ-saturated structures).

PROPOSITION 3.4 Let $At \cup \neg\Psi \subseteq \Psi$, K ∞-compact. Then t.f.a.e. :
(i) For all $\underline{A} \overset{\Psi}{\preccurlyeq} \underline{B}$ in K and all finite $A' \subseteq A$, $B' \subseteq B$ there exists $\underline{A} \preccurlyeq \underline{C} \in K$ and a Ψ-preserving morphism $\underline{B} \supseteq A' \cup B' \xrightarrow{f} \underline{C}$ with $f|A' = id$.
(ii) K is Ψ-model-complete.
(iii) For all $\underline{A} \overset{\Psi}{\preccurlyeq} \underline{B}$ in K there exists $\underline{B} \overset{\Psi}{\preccurlyeq} \underline{C} \in K$ with $\underline{A} \preccurlyeq \underline{C}$.

PROPOSITION 3.5 Let $At \cup \neg\Psi \subseteq \Psi$, K ∞-compact. Then t.f.a.e. :
(i) K is Ψ-substructure-complete .
(ii) For all $\underline{A},\underline{B} \in K$, $A' \subseteq A \cap B$ with $\underline{A} \overset{\Psi}{\equiv}_{A'} \underline{B}$ there exists $\underline{A} \preccurlyeq \underline{C} \in K$ and a Ψ-preserving morphism $\underline{B} \xrightarrow{f} \underline{C}$ with $f| A' = id$.
(iii) For all $\underline{A},\underline{B} \in K$, all finite $A' \subseteq A \cap B$ with $\underline{A} \overset{\Psi}{\equiv}_{A'} \underline{B}$ and all $b \in B$ there ex. $\underline{A} \preccurlyeq \underline{C} \in K$ and a Ψ-preserving morphism $\underline{B} \supseteq A' \cup \{b\} \xrightarrow{f} \underline{C}$ with $f| A' = id$.

PROOF. In both 3.4 and 3.5, (ii) $\Longrightarrow$ (iii) $\Longrightarrow$ (i) is trivial. For (i) $\Longrightarrow$ (ii) observe that by ∞-compactness $Diag_{Fo}(\underline{A},A) \cup Diag_{\Psi}(\underline{B},B)$ has a model $\underline{D}$ with $\underline{D}|L \in K$ (in 3.5 assume w.l.o.g. $A' = A \cap B$).

ADDENDUM 3.6 Using the basic properties of κ-saturated structures and ultrapowers, 3.4 and 3.5 can be modified as follows: 3.4(iii) can be replaced by 3.4(iii'): For all $\underline{A} \overset{\Psi}{\preccurlyeq} \underline{B}$ in K and all $card(B)^+$-saturated $\underline{A} \preccurlyeq \underline{C} \in K$ there exists a Ψ-preserving morphism $\underline{B} \xrightarrow{f} \underline{C}$ with $f|A = id$. In 3.5 the phrase " there exists ..." can be replaced by "and for all $card(B)^+$-saturated $\underline{A} \preccurlyeq \underline{C} \in K$ there exists a Ψ-preserving morphism $\underline{B} \xrightarrow{f} \underline{C}$ with $f|A' = id$."Similarly in 3.4(i) and 3.5(iii): Here the phrase "there exists $\underline{A} \preccurlyeq \underline{C} \in K$ ..." can be replaced by "for all ω-saturated $\underline{A} \preccurlyeq \underline{C} \in K$ ..."; for elementary K also by "for all non-principal ultrapowers $\underline{C} = \underline{A}^{\omega}/F$ of $\underline{A}$...", provided L is countable, cf. [ChKe 1973, thm.6.1.1.] Finally, one may require instead " $\underline{A} = \underline{C}$ ω-saturated ".

As a consequence of 3.6 and 3.5 for Ψ = Ba, we get:

COROLLARY 3.7 Let K be ∞-compact and At-complete. Then K admits q.e. iff every ω-saturated $\underline{A} \in K$ is ω-homogeneous and ω-universal in K .

§ 4 CHARACTERIZATION OF Q.E.-STRUCTURES .

$\underline{A}$ is a q.e.-structure if $\underline{A}$ (or equivalently $\{\underline{B} : \underline{B} \equiv \underline{A}\}$) admits q.e.. The methods described in sections 1-3 have been applied to many types of algebraic L-structures $\underline{A}$ to prove q.e. in $\underline{A}$, where L is a natural, algebraic language. Examples include: Fields (cf. AdKi 1976,Me 1979,Ro 1963,Sh 1976,Wo 1979,Wr 1983), ordered fields (cf. Dr 1978,KrKr 1967,Pl 1981/82, Se 1954,Sh 1977,TaMK 1951), valued fields (cf. AxKo 1966,Bs 1979,Co 1969,De , Er 1967/80,Ma 1976,Ro 1956, We 1971/76/c,Zi 1972), differential fields (cf. Ro 1963,Sa 1972,Se 1956a,PlRq 1984 Wo 1973/76), commutative rings (cf. BoCh 1980,Ca 1973,ChDi 1980,LpSr 1973,Ma 1973, We 1975/82), (ordered) abelian groups & modules (cf. EkFi 1972,EkSb 1971,Ba 1976, Kr 1975,Mk ,Scht ,We 1981/a,Zi), distributive lattices (cf. Kr 1975,KrKr 1967, Schd 1982, Scht 1976).

As reaction, model theorists began in the mid-197o's to ask themselves, whether these were all q.e. structures in the given algebraic class. The outcome can roughly be classified into three cases :

Case 1 : The known q.e. structures are the only ones. Examples: fields (MaKD 1983, Wh 1979), ordered fields & rings (MaKD 1983, Dr 198o, Wh 1979) , rings of char. o (Rs 1978,Be 1981) , valued fields (MaKD 1983), ord. & l.o. ab. groups (cf. Po 1983), distr. lattices (We b), Stone algebras (Ra).

Case 2 : q.e. structures can be completely described from known examples by some constructions (e.g. direct products, boolean products). Examples: ab. groups (Bl 1981, ChFe 1982, We a), semisimple rings (BoMP 198o), biregular rings (Po 1981), distr. lattice with constants (We b).

Case 3 : There is a huge number of q.e. structures for which a stringent, but not quite complete description is available. Examples: groups (ChFe 1982, Fg , SrWo 1982) , rings (Be 1980,1981 , BeCh 1981,1981a,1983 , SrWo).

The case of modules does not seem to be sufficiently clarified (cf. Pr 1981,We a).

Another recent development is the characterization of structures that allow reduction of an algebraically interesting set Φ of formulas to Q_o (e.g."linear" formulas in rings, Dr 1981, Po 1983, We c).

In order to show that a set Φ of formulas is not reducible to Q_o in $\underline{A}$, one employs as a rule definable sets and relations: It suffices to find tuples $\underline{a},\underline{b}$ from A that satisfy the same atomic formulas in $\underline{A}$, and a formula $\varphi(\underline{x}) \in \Phi$

with $\underline{A} \models \varphi(\underline{a}) \wedge \neg\varphi(\underline{b})$.

In the following, we outline some of these results. In each case, the outcome is extremely sensitive to a change of language: Too small a language may yield only trivial q.e. structures (consider e.g. bounded chains in the language $\{<\}$); an extension of the language by algebraically natural (preferably definable) constants, operations and relations may greatly increase the types of q.e. structures in the given class. This fact creates a wealth of algebraically significant problems.

4.1 Fields and rings. Let $L = \{0,1,+,-,\cdot\}$. Any finite field and any algebraically closed field admits q.e. in L. Conversely, any ring of characteristic 0, any division ring, any prime ring with infinite center that admits q.e. in L is of this type. Any q.e. ring without nilpotent elements is a finite direct product of q.e. fields and atomless p^n-rings. (A p^n-ring is a commutative ring satisfying the identity $x^{p^n} = x$ and containing the finite field $\mathbb{F}_{p^n}$; "atomless" refers to the boolean algebra B(R) of idempotents of R.) Any q.e. semisimple ring is a finite product of q.e. fields, atomless p^n-rings and 2×2-matrix rings $M_2(\mathbb{F}_p)$. An infinite integral domain admits q.e. for linear formulas iff it is a field.

Problems. (i) Let D(x,y) be defined as "x divides y", let $L_D = L \cup \{D\}$. Characterize all commutative rings that admit q.e. in L_D. (Monically closed valuation rings are examples of this kind that are not fields. cf. [We 1982, c])
(ii) Let C be a set of new constants and let $L_C = L \cup C$. Characterize all rings that admit q.e. in L_C; similarly for $L_{D,C}$.
(iii) Let $L_Z = L \cup \{Z_n\}_{1<n<\omega}$, where $Z_n(x_0,\dots,x_n)$ is defined by $\exists y(\sum_{0\leq i\leq n} x_i y^i = 0)$. Characterize all fields that admit q.e. in L_Z. (Examples that are not algebraically closed are the pseudofinite fields. cf. AdKi 1976 , more generally ChDM)
(iv) Let $L_W = L \cup \{W_n\}_{1<n<\omega}$, where $W_n(x)$ is defined by $\exists y(y^n = x)$. Are there non-algebraically closed, infinite fields that admit q.e. in L_W ?

4.2 Ordered rings. A linearly ordered ring admits q.e. in $L_< = L \cup \{<\}$ iff it is a real-closed field.
Problem. Characterize q.e. rings in $L_{<,D}$ and $L_{<,D,C}$ for a set C of constants. (Real-closed rings are examples that are not fields. cf. [ChDi 1980])

4.3 Valued fields. Let $L_{pf} = L \cup \{V\} \cup \{W_n\}_{1<n<\omega}$, where V is a predicate for the valuation ring and W_n are as in 4.1(iv). A p-field is a L_{pf}-substructure of a p-adically closed field. Then a p-field admits q.e. in L_{pf} iff it is p-adically closed. Let $L_{vf} = L \cup \{D\}$, where D is a linear divisibility relation (cf. MaKD 1983). A nontrivially valued field admits q.e. in L_{vf} iff it is algebraically closed. [De] contains a general theorem on relative q.e. in the equal characteristic case.

4.4 Differential fields. Let $L' = L \cup \{'\}$,where ' is the operation of derivation. Any differentially closed field of characteristic 0 admits q.e. in L'. Let $L'(r) = L' \cup \{r\}$,where r is an operation assigning to each x with $x' = 0$ a p-th root. Any differentially closed field of characteristic $p \neq 0$ admits q.e. in L'(r).
Problem. Find a converse to these results.

4.5 R-Modules. Let $L(R) = \{0,+,-,\{r\cdot\}_{r \in R}\}$. In any R-module all formulas are equivalent to boolean combinations of positive primitive formulas. R is von Neumann regular iff every R-module admits q.e.. Let R be a Dedekind domain. Then an R-module $\underline{M}$ admits q.e. iff $\underline{M}$ is divisible or M is a torsion module with only finitely many nonzero primary submodules $\underline{M}_p$ such that each $\underline{M}_p$ is a direct sum of copies of a fixed cyclic module R/P^n. This applies in particular to q.e. abelian groups.
Problem. Characterize q.e. abelian groups with distinguished subgroups.

4.6 Ordered abelian groups. Let $L_{og} = \{0,+,-,<\}$. An ordered abelian group admits q.e. in L_{og} iff it is divisible. In $L_{og}^{\equiv} = L_{og} \cup \{\equiv_n\}_{1<n<\omega}$, where $x \equiv_n y \leftrightarrow \exists z(\, nz = x-y\,)$, the q.e. ordered abelian groups are exactly the dense regular groups. In $L_{og}^{\equiv} \cup \{1\}$, a discrete abelian group with smallest positive element 1 admits q.e. iff it is a Z-group.
Problem. Characterize q.e. ord. ab. groups in $L_{og}^{\equiv} \cup C$ for a set C of constants.

4.7 Lattices. A distributive lattice admits q.e. in $L_{\mathcal{L}} = \{\sqcap,\sqcup\}$ iff it is trivial, a chain or a dense, relatively complemented lattice without endpoints. A boolean algebra admits q.e. in $L_{\mathcal{L}} \cup \{0,1\}$ iff it is atomless or has 1,2, or 4 elements. Distributive lattices that admit q.e. in $L_{\mathcal{L}} \cup C$ for a set C of constants are completely characterized in [We b] . Stone algebras that admit q.e. in $L_{\mathcal{L}} \cup \{0,1,*\}$ are classified in [Ra] .
Problems. Characterize all q.e. non-distributive lattices in $L_{\mathcal{L}} \cup C$, all q.e. distributive p-algebras in $L_{\mathcal{L}} \cup \{0,1,*\}$, all q.e. Heyting algebras in $L_{\mathcal{L}} \cup \{0,1,\rightarrow\}$.

4.8 Boolean products and applications. Let L contain a constant c, and let L^* be the associated language of boolean products of L-structures , P the theory of boolean products in L^* (cf. We 1975). Then a model $\underline{A}$ of P admits q.e.

in L^* iff there exist quantifier-free L-sentences $\varrho_1,\ldots,\varrho_n$ such that:

(i) The truth values $v\varrho_1,\ldots,v\varrho_n$ form a partition of 1 in the boolean algebra $\underline{A}_{\mathcal{B}}$, where $\underline{A} = (\underline{A}_L, \underline{A}_{\mathcal{B}}, v_=, v_R, \ldots)$;

(ii) The classes $\{\underline{A}_i : i \in I, v\varrho_k \notin i\}$ of canonical stalks above $v\varrho_k$ admit q.e. in L. ($I = \mathrm{spec}(\underline{A}_{\mathcal{B}})$ is the set of prime ideals of $\underline{A}_{\mathcal{B}}$.)

(iii) The intervals $[0, v\varrho_k]$ have 1,2, or 4 elements or are atomless.

This theorem can be applied in combination with 4.1, 4.2, 4.6 to characterize:

(i) All commutative von Neumann regular rings that admit q.e. in the language $\{0,1,+,-,\cdot,*\}$, where $1-a^*$ is the idempotent generating Ann(a).

(ii) All commutative function rings with 1 and projector that admit q.e. in $\{0,1,+,-,\cdot,\sqcap,\sqcup,pr\}$.

(iii) All lattice-ordered abelian groups with projector and weak unit e that admit q.e. in $\{0,+,-,\sqcap,\sqcup,pr,e\}$.

In [Po 1983] generalizations of (i)-(iii) (characterization of q.e. biregular rings and q.e. abelian lattice-ordered groups with projector in appropriate languages) have been obtained by a different method.

Finally, we mention that even explicit primitive recursive q.e. procedures (and decision procedures) are in general far from being feasable on a computer. Therefore, a great deal of effort has been spent on finding procedures that are at least elementary-recursive, i.e. whose time complexity is bounded by a fixed finite iteration of the exponential function 2^n (cf. Bö 1983, Br 1978, Cl 1983, Cp 1972, FrRk 1975, Mk 1975, Mk).Here one basic problem is to avoid the use of criterion 1.2(ii),which involves the formation of disjunctive normal forms, by restricting quantifiers to finite, effectively described systems of representatives.

For applications of quantifier elimination to Nullstellensätze in fields, rings, differential fields and rings, to definite functions in ordered fields, to integer-valued functions in valued fields, and the like we refer the reader e.g. to (Dr 1979, Mc 1980, PlRq 1984, Ro 1963, Se 1956/56a, We 1975a/77/82) and the references given in these articles.

BIBLIOGRAPHY . It is a futile enterprise to try to give a reasonably complete bibliography on quantifier elimination (including model-completeness) on a few pages. So the references provided below are selected not only for their own sake, but also as source of further references. Besides, the reader should consult the references in the Handbook of Mathematical Logic, North-Holland, Amsterdam, 1977.

AdKi 1976 A.Adler,C.Kiefe, Pseudofinite fields, procyclic fields & model-completion, Pac.J.Math. 62, 305-309.

AxKo 1966 J.Ax,S.Kochen, Diophantine problems over local fields III, Ann.Math. 83, 437-456.

Bc 1973 P.D.Bacsich, Primality & model completions, Alg.Univ. 3, 265-270.

BnLn 1973 J.T.Baldwin,A.H.Lachlan, On universal Horn classes categorical in some infinite power, Alg.Univ. 3, 98-111.

Bs 1979 S.A.Baserab, A model-theoretic transfer theorem for henselian valued fields, J.reine & angew. Math. 311/312, 1-30.

Bd 1975 A.Baudisch, Die elem. Theorie der Gruppe vom Typ p^{∞} mit Untergruppen, Z.Math. Logik & Grundl. Math. 21, 347-352.

Ba 1976 W.Baur, Elimination of quantifiers for modules, Israel J. Math. 25, 64-70.

Bl 1981 O.V.Belegradek, Abelian groups which admit e.q.., Abstracts AMS, 81T-03-489.

Be 1980 C.Berline, Elim. of quant. for non semi-simple rings of char. $p \neq 0$, Model Th. of Alg. & Arithm. , Proc. Karpacz, Springer LNM vol. 834.

Be 1981 -- " -- , Rings which admit elim. of quantifiers, J. Symb. Logic 46, 56-58.

BeCh 1981 C.Berline,G.Cherlin, QE rings in characteristic p, Logic Year 1979-80, Springer LNM, vol.859.

BeCh 1981a -- " -- , QE nilrings of prime characteristic, Bull.Soc.Math.Belg. 23, 3-17.

BeCh 1983 -- " -- , QE rings in characteristic p^n, J. Symb. Logic 48, 140-162.

Bo 1980 M.Boffa, Elimination des quantificateurs en algebre, Bull.Soc.Math.Belg. 22, 107-133.

BoMP 1980 M.Boffa,A.Macintyre,F.Point, The q.e.-problem for rings without nilpotent elements & for semisimple rings, Model th. of Alg. & Arithm., Proc. Karpacz, Springer LNM vol. 834.

BoCh 1980 M.Boffa,G.Cherlin, Elim. des quantificateurs dans les faisceaux, C.R.Acad.Sc. Paris 290, 355-357.

Bö 1983 W.Böge, Quantifier elimination for the elementary real algebra, manuscript, Heidelberg.

Br 1978 S.S.Brown, Bounds on transfer principles for algebr. closed & complete discretely valued fields, Mem. AMS 204 .

Bu 1975 S.Burris, An existence theorem for model companions, Proc. Conf. Lattice theory, Ulm 1975.

Ca 1973 A.B.Carson, The model completion of the theory of commutative regular rings, J. Algebra 27, 136-146.

Ce 1980 P.Cegielski, La théorie élémentaire de la multiplication, C.R.Acad.Sc.Paris, 290, sér.A, 935-938.

ChKe 1973 C.C.Chang,H.J.Keisler, Model Theory, North-Holland, Amsterdam.

ChDi 1980 G.Cherlin,M.Dickmann, Anneaux réel-clos et anneaux des fonctions continues, C.R.Acad.Sc. Paris 290, 1-4.

ChDM --- G.Cherlin,L.van den Dries,A.Macintyre, The elementary theory of regularly closed fields, preprint.

ChFe 1982 G.Cherlin,U.Felgner, Quantifier eliminable groups, Logic Coll.'80, North-Holland, Amsterdam, 69-81.

Co 1969 P.Cohen, Decision procedures for real and p-adic fields, Comm. pure & appl. Math. 22, 131-153.

Cl 1983 G.E.Collins, QE for real closed fields; a guide to the literature, Computer Algebra,Symb. & Alg. Computation, 2nd ed.,Springer, Wien.

Cm 1976 S.D.Comer, Complete & model-complete theories of monadic algebras, Coll. Math. 34, 183-190.

Cp 1972 D.C.Cooper, Theorem proving in arithmetic without multiplication, Machine Intelligence vol.7 , 91-99.

De --- F.Delon, Quelques proprietés des corps values en théorie des modeles, these, Paris 7.

Dr 1978 L.van den Dries, Model theory of fields, thesis, Utrecht.

Dr 1979 -- " -- , Artin-Schreier theory of comm. regular rings, Ann. math. Logic 12, 113-150.

Dr 1980 -- " -- , A linearly ordered ring whose theory admits q.e. is a real closed field, Proc. AMS 79, 97-100.

Dr 1981 -- " -- , QE for linear formulas over ordered & valued fields, Bull.Soc.Math.Belg. 23, 19-32.

Dr 1982 -- " -- , Some applications of a model theoretic fact..., Indag. Math. 44, 397-401.

Dr a -- " -- , Remarks on Tarski's problem concerning (R,+, ,exp), to appear in Logic Coll.'82, Florence.

Ek 1972 P.Eklof, Some model theory of abelian groups, J.Symb.Logic 37, 335-342.

EkFi 1972 P.Eklof,E.Fisher, The elementary theory of abelian groups, Ann. math. Logic 4, 115-171.

EkSb 1971 P.Eklof,G.Sabbagh, Model completions & modules, Ann. math. Logic 2, 251-295.

Er 1967 Y.Ershov, Fields with a solvable theory, Sov. Math.Dokl. 8, 575-576.

Er 1980 -- " -- , Multiply valued fields, Sov. Math. Dokl. 22, 63-66.

ELTT 1965 Y.Ershov,I.A.Lavrov,A.D.Taimanov,M.A.Taitslin, Elementary theories, Russian Math. Surveys 20, 4, 35-105.

FeVa 1959 S.Feferman,R.Vaught, The first-order properties of products..., Fund. Math. 47, 57-103.

Fg --- U.Felgner, The classification of all quantifier-eliminable FC-groups, preprint.

FrRk 1975 J.Ferrante,C.Rackoff, A decision procedure for the first order theory of real addition with order, SIAM j. Comput. 4, 69-76.

Fl 1976 I.Fleischer, QE for modules & ordered groups, Bull. Acad. Polon. Sc. 24, 9-15.

FrSc 1976 M.Fried,G.Sacerdote, Solving diophantine problems over all residue class fields..., Ann. Math. 104, 203-233.

Ga 1980 S.Garavaglia, Decomposition of totally transcendental modules, J. Symb. Logic 45, 155-164.

GlPi 1980 A.Glass,K.Pierce, Existentially complete abelian lattice-ordered groups, Trans. AMS 261, 255-270.

Gu 1977 Y.Gurevich, Expanded theory of ordered abelian groups, Ann. math. Logic 12, 193-228.

HiWh 1975 J.Hirschfeld,W.Wheeler, Forcing, arithmetic, division rings, Springer LNM vol. 454.

Ho 1980 W.Hodges, Functorial uniform reducibility, Fund. Math. 108, 77-81.

Ja 1980 B.Jacob, The model theory of "R-formal" fields, Ann. math. Logic 19, 263-282.

JdKn 1975 M.Jarden,U.Kiehne, The elementary theory of algebraic fields of finite corank, Inv. Math. 30, 275-294.

JdRq 1980 M.Jarden,P.Roquette, The Nullstellensatz over the p-adic numbers, J.Math.Soc. Japan 32, 425-460.

Ko 1962 S.Kochen, Ultraproducts in the theory of models, Ann. Math. 74, 221-261.

Ko 1975 -- " -- , The model theory of local fields, Logic Conf. Kiel 1974, Springer LNM vol. 499.

KvKn 1969 G.T.Kozlov,A.I.Kokorin, Elementary theory of abelian groups without torsion, with a predicate selecting a subgroup, Algebra & Logic 8, 182-190.

Kr 1975 P.Krauss, Quantifier elimination, Logic Conf. Kiel 1974, Springer LNM vol. 499.

Kr 1977 -- " -- , Homogeneous universal models of universal theories, Z.math. Logik & Grundl. Math. 23, 415-426.

KrKr 1967 G.Kreisel,J.L.Krivine, Elements of mathematical Logic, North-Holland, Amsterdam.

La 1927 C.H.Langford, Some theorems on deducibility, Ann.Math. 28, 459-471.

Li 1964 P.Lindström, On model-completeness, Theoria 30, 183-196.

LpSr 1973 L.Lipshitz,D.Saracino, The model companion of the theory of comm. rings without nilpotent elements, Proc. AMS 38, 381-387.

Ma 1973 A.Macintyre, Model-completeness for sheaves of structures, Fund. Math. 81, 73-89.

Ma 1976 -- " -- , On definable sets of p-adic numbers, J. Symb. Logic 41, 605-610.

Ma 1977 -- " -- , Model completeness , in Handbook of mathematical logic, North-Holland, Amsterdam, 139-180.

MaKD 1983 A.Macintyre,K.McKenna,L.v.d.Dries, Elimination of quantifiers in algebraic structures, Adv. in Math. 47, 74-87.

Mv 1971 A.I.Malcev, Axiomatic classes of locally free algebras, in Metamathematics of algebraic systems, North-Holland, Amsterdam.

Mn 1980 K.Manders, Theories with the existential substructure property, Z. math. Logik & Grundl. Math. 26, 98-92.

Mc 1980 K.McKenna, Some diophantine Nullstellensätze, Model Theory of Algebra & Arithm., Proc. Karpacz, Springer LNM vol. 834.

Mz 1982 H.C.Mez, Existentially closed linear groups, J. Algebra 76, 84-98.

Mk 1975 L.Monk, Elementary-recursive decision procedures, Ph.D. dissertation, Univ. California, Berkeley.

Mk a --"-- , The theory of abelian groups is Kalmar-elementary, preprint.

MrVa 1962 M.Morley,R.Vaught, Homogeneous universal models, Math.Scand.11, 37-57.

MoTa 1949 A.Mostowski,A.Tarski, Arithmetical classes & types of well-ordered systems, Bull. AMS 55, 65.

MyPm 1981 J.Mycielski,P.Perlmutter, Model-completeness of some metric completions of absolutely free algebras, Alg. Univ. 12, 137-144.

Pa 1983 M.Parigot, Le modèle compagnon de la théorie des arbres, Z. math. Logik & Grundl. Math. 29, 137-150.

Po 1981 F.Point, Elimination des quantificateurs dans les L-anneaux -reguliers, Bull.Soc.Math.Belg. 23, 93-108.

Po 1983 --"-- , QE for projectable L-groups and linear elimination for rings, These, Mons.

Ph 1981 K.Potthoff, Einführung in die Modelltheorie & ihre Anwendungen, wiss. Buchgesellschaft, Darmstadt.

Pb 1929 M.Presburger, Über die Vollständigkeit eines gewissen Systems der Arithmetik..., C.R. 1er Congr. des pays slaves,Warsaw, 92-101.

Pr 1981 M.Prest, QE for modules, Bull.Soc.Math.Belg. 23, 109-130.

Pl 1981 A.Prestel, Pseudo real-closed fields, Set theory & model theory, Proc. Bonn 1979, Springer LNM vol.872.

Pl 1982 -- " --, Decidable theories of preordered fields, Math. Ann. 258, 481-492.

PlRq 1984 A.Prestel,P.Roquette, Lectures on formally p-adic fields, Springer LNM.

Ra --- S.Rauschning, QE for Stone algebras, Diplomarbeit,Heidelberg, in preparation.

Ro 1956 A.Robinson, Complete theories, North-Holland, Amsterdam.

Ro 1963 -- " -- , Introduction to model theory & the metamathematics of algebra, North-Holland, Amsterdam.

Rs 1978 B.Rose, Rings which admit elimination of quantifiers, J. Symb. Logic 43, 92-112.

Rs 1980 --"-- , Prime q.e. rings, J.London Math. Soc. 21, 257-262.

Sa 1972 G.Sacks, Saturated Model Theory, Benjamin, Reading Mass.

Sr 1973 D.Saracino, Model companions for $\aleph_0$-categorical theories, Proc. AMS 39, 591-598.

SrWo 1982 D.Saracino,C.Wood, QE nil-2-groups of exponent 4, J.Alg. 76, 337-352.

SrWo --- -- " -- , QE comm. nilrings, to appear in J.Symb.Logic.

Schd 1982 J.Schmid, Model companions of distr. p-algebras, J. Symb. Logic 42, 680-688.

Scht 1976 P.Schmitt, The model completion of Stone algebras, Ann.Sci.Univ.Clermont Math. 13, 135-155.

Scht --- --"-- , Model- & substructure-complete theories of ordered abelian groups, to appear in Logic Coll.'83,Aachen.

SSmT 1983 W.Schwabhäuser,W.Szmielew,A.Tarski, Metamathematische Methoden in der Geometrie , Springer Verlag.

Se 1954 A.Seidenberg, A new decision method for elementary algebra, Ann. Math. 60, 365-374.

Se 1956 -- " -- , Some remarks on Hilbert's Nullstellensatz, Arch. Math. 7, 235-340.

Se 1956a -- " -- , An elimination theory for differential algebra, Univ. California. Pzbl. 3, 31-66.

Sh 1967 J.Shoenfield, Mathematical logic, Addison-Wesley,Reading,Mass.

Sh 1971 -- " -- , A theorem on quantifier elimination, Symposia Math., vol. V, Rome 1969/70, Acad. Press, London.

Sh 1976 -- " -- , Quantifier elimination in fields, Non-class. Logics, Model Theory & Comp.,Proc. Campinas 1976, North-Holland, Amsterdam.

Si 1978 M.Singer, The model theory of ordered differential fields, J. Symb. Logic 43, 82-91.

Sko 1919 Th.Skolem, Untersuchungen über die Axiome des Klassenkalküls... .

Sko 1920 -- " -- , Logisch-kombinatorische Untersuchungen über die Erfüllbarkeit..

Sko 1928 -- " -- , Über die mathematische Logik.

Sko 1930 -- " -- , Über einige Satzfunktionen in der Arithmetik. All articles in Th.Skolem, Selected works in logic (J.E.Fenstad Ed.) , Universitetsforlaget, Oslo 1970.

St 1976 W.Stegbauer, A generalized model companion for a theory of partially ordered fields, J. Symb. Logic 44, 643-652.

Sm 1949 W.Szmielew, Arithmetical classes & types of abelian groups, Bull. AMS 55, 65.

Sm 1955 -- " -- , Elementary properties of abelian groups, Fund. Math. 41, 203-271.

Ta 1949 A.Tarski, Arithmetical classes & types of boolean algebras, Bull. AMS 55, 64.

Ta 1949a -- " -- , Arithmetical classes & types of algebraically closed & real-closed fields, Bull.AMS 55, 64.

TaMK 1951 A.Tarski,J.C.C.McKinsey,A decision method for elementary algebra & geometry, Berkeley .

To 1975 C.Toffalori, Eliminazione dei quantificatori per certe teorie di coppie di campi, Boll. Un. Mat. Ital. A 15, 159-166.

Tr 1979 R.Transier, Verallgemeinerte formal p-adische Körper, Arch. Math. 32, 572-584.

Tu 1981 S.Tulipani, Model-completeness of theories of finitely additive measures with values in an ordered field, Z. math. Logik & Grundl. Math. 27, 481-488.

Vo 1976 H.Volger, The Feferman-Vaught theorem revisited, Coll.Math. 36, 1-11.

Wa 1931 B.L.van der Waerden, Moderne Algebra, Kapitel 11, Springer, Berlin.

Wb 1898 H.Weber, Lehrbuch der Algebra, Abschnitt 4, reprint Chelsea, N.Y..

We 1971 V.Weispfenning, Elementary theories of valued fields, doctoral dissertation, Heidelberg.

We 1975 --- " --- , Model-completeness and elimination of quantifiers for subdirect products of structures, J. Algebra 36, 252-277.

We 1975a --- " --- , Two model theoretic proofs of Rückert's Nullstellensatz, Trans. AMS 203, 331-342.

We 1976 --- " --- , On the elementary theory of Hensel fields, Ann. math. Logic 10, 59-93.

We 1977 --- " --- , Nullstellensätze - a model theoretic framework, Z. math. Logik & Grundl. Math. 23, 539-545.

We 1978 --- " --- , A note on $\aleph_0$-categorical model companions, Arch. math. Logik 19, 23-29.

We 1981 --- " --- , QE for certain ordered & lattice-ordered abelian groups, Bull.Soc.Math.Belg. 23, 131-156.

We 1982 --- " --- , Valuation rings & boolean products, Proc. Conf. F.N.R.S., Brussels.

We a --- " --- , QE for modules, to appear in Arch. math. Logik.

We b --- " --- , QE for distributive lattices & measure algebras, to appear in Z. math. Logik & Grundl. Math.

We c --- " --- , QE & decision procedures for valued fields,preprint.

Wr 1983 W.Wernecke, Über die elementare Theorie separabel abgeschlossener Körper, Diplomarbeit, Heidelberg.

Wh 1978 W.Wheeler, A characterization of companionable, universal theories, J. Symb. Logic 43, 402-429.

Wh 1979 --- " ---, Amalgamation & quantifier elimination for theories of fields, Proc. AMS 77, 243-250.

Wh 1979a --- " ---, Model-complete theories of PAC-fields, Ann. math. Logic 17, 205-226.

Wh 1980 --- " ---, Model theory of strictly upper triangular matrix rings, J. Symb. Logic 45, 455-463.

Wh --- --- " ---, ω-homogeneous universal rings, preprint.

Wi 1975 P.Winkler, Model-completeness & Skolem expansions, in Model Theory & Algebra, Springer LNM vol. 498.

Wo 1973 C.Wood, The model theory of differential fields of characteristic $p \neq 0$, Proc. AMS 40, 577-584.

Wo 1976 --"-- , The model theory of differential fields revisited, Israel J. Math. 25, 331-352.

Wo 1979 --"-- , Notes on the stability of separably closed fields, J. Symb. Logic 44, 412-416.

Za 1974 E.Zakon, Model-completeness & elementary properties of torsion-free abelian groups, Canad. J. Math. 26, 829-840.

Zi 1972 M.Ziegler, Die elementare Theorie henselscher Körper, doctoral dissertation, Köln.

Zi --- -- " -- , Model theory of modules, preprint.

Universal Algebra and its Links with Logic,
Algebra, Combinatorics, and Computer Science
Proc. "25. Arbeitstagung über Allgemeine Algebra", Darmstadt 1983
P. Burmeister et al. (eds.)

106 – 114

BOOLEAN CONSTRUCTIONS AND THEIR RÔLE IN UNIVERSAL ALGEBRA AND MODEL THEORY

H. Werner

In recent developments in universal algebra and model theory the notion of Boolean power and its generalizations became very important and the reader will find a very detailed exposition of these notions in Burris & Sankapanavar [4] where he also can find an excellent bibliography on this subject.

Boolean powers originate from the work of Ahrens and Kaplansky in 1948, who were interested in topological representations of rings, in particular they were interested in the ring of all continuous functions from some topological space into some ring endowed with the discrete topology. As it turns out immediately the topological space in this situation can always be chosen to be boolean, i.e. compact with a basis of clopen (= closed and open) sets. Obviously inspired by this construction A.L. Foster in 1953 invented the notion of Boolean extension, which nowadays is called Boolean power:

If A is a finite algebra and B a Boolean algebra, then
$$A[B] := \{f : A \longrightarrow B \mid \bigvee_{a\in A} f(a) = 1 \;\&\; a \neq b \Rightarrow f(a) \wedge f(b) = 0\}$$
and the operations on $A[B]$ are defined similarly as:
$$(f + g)(a) := \bigvee_{b+c=a} f(b) \wedge f(c)$$, where $+$ is an operation on A .

Soon B.Jónsson observed that the two notions are the same, namely the map assigning to each continuous map $f : X \longrightarrow A$ * the map $\overline{f} : A \longrightarrow B$ $B = \{U \subseteq X \mid U \text{ clopen}\}$ with $\overline{f}(a) = \{x \in X \mid f(x) = a\}$ is an isomorphism. Finally 1975 S.Burris gave a description of Boolean powers as special subdirect products, which is very close to the set-theoretic notion of boolean-valued structures, and this notion allowed lots of generalizations to more complex situations. Burris's description reads as follows:

* X denotes the Stone-space of B and hence B can be considered to be the Boolean algebra of all clopen subsets of X .

$A[B] \simeq C(X,A) \hookrightarrow A^X$ is a subdirect product such that

(i) For all $a,b \in A[B]$ the <u>equalizer</u>
$[a = b] = \{x \in X \mid a(x) = b(x)\}$
is a clopen subset of X (<u>Equalizer condition</u> (EC))

(ii) <u>Patchwork-property (PP)</u>:
For each clopen subset $N \subseteq X$ and $a,b \in A[B]$
also $a\restriction N \cup b\restriction X\setminus N$ belongs to $A[B]$

(iii) All constants belong to $A[B]$

Some further properties of $A[B]$ will be of importance later:

(iv) <u>Maximum-property (MP)</u>:
For each formula $\varphi(x)$ (with constants from $A[B]$)
the set of all <u>solution sets</u>
$[\varphi(a)] = \{x \in X \mid A \models \varphi(a(x))\}$ has a maximum,

or equivalently:

For each sentence φ (with constants from $A[B]$)
the solution set $[\varphi]$ is clopen.

(v) We can specialize (PP) by demanding N to be an equalizer:
(P_o) : $\forall a,b,c,d \in A[B]\ \exists u \in A[B]$
$[c = d] \subseteq [a = u]\ \&\ [c \neq d] \subseteq [b = u]$
and this condition is satisfied iff $A[B] \subseteq A^X$
is closed under the <u>ternary discriminator</u> t on A

$$t(x,y,z) = \begin{cases} x & \text{if } x \neq y \\ z & \text{if } x = y \end{cases} \quad \text{for all } x,y,z \in A .$$

This last observation shows that there are many varieties of algebras such that each member is a Boolean power of some algebra A (namely if A has no subalgebras and carries the ternary discriminator).

The widest generalization of Boolean powers is the notion of a <u>Boolean product</u> or <u>global subdirect product</u>, which is a subdirect product $R \hookrightarrow \prod_{x\in X} A_x$ such that X carries a boolean topology and R satisfies the equalizer condition and the patchwork property with respect to that topology. Since this notion is equivalent to the

notion of an algebra of all continuous global sections of a sheaf with base space X and stalks A_x $(x \in X)$ we refer to the algebras A_x as to the stalks of the Boolean product (global subdirect product) R . For a class $\mathbb{A}$ of algebras $\Gamma^a(\mathbb{A})$ denotes the class of all Boolean products with stalks in $\mathbb{A}$ and $\Gamma^e(\mathbb{A})$ denotes the subclass of those Boolean products which also satisfy the maximum property (MP).
In the book "Discriminator algebras" [15] the author gives many examples of varieties which are of the form $\Gamma^a(\mathbb{A})$ or even $\Gamma^e(\mathbb{A})$.
Important special cases of Boolean products are

(1) sub-Boolean-powers (filtered Boolean powers) of A :

i.e. subalgebras of A[B] satisfying (PP).

(2) Boolean powers with homeomorphisms:

Subalgebras of A[B] preserving the action of Aut(A) or of some distinguished subgroup G of Aut(A) .

(3) Double-Boolean-powers:

Subalgebras of A[B] defined by a distinguished subalgebra C of B and a congruence θ on A as follows

$$R = \{f \in A[B] \mid f^{-1}(a/_{\theta}) = \bigvee_{b\theta a} f^{-1}(b) \in C\}$$

After having outlined the basic notions of Boolean constructions (for more details the reader is referred to [2]) we want to give a short review of some of the development inspired by these notions and connected to model theoretic questions. The material presented here is mostly connected to the names S.Bulman-Fleming, S. Burris, D.Clark, S.Comer, B.A.Davey, D.Haley, T.K.Hu, K.Keimel, P.Krauss, R.McKenzie, A.F.Pixley, R.W.Quackenbush, H.Volger, and A.Wolf.

We restrict our attention to discriminator varieties, although several results could be pushed further e.g. to filtral varieties (P.Krauss) or congruence-modular varieties (S.Burris & R.McKenzie).

A discriminator variety is a variety V of algebras (and in this review we always assume finite type) having a term t, which is the ternary discriminator on all subdirectly irreducible members of V. In this case the class A of all subd.irred. members of V is a universal axiomatic class and $V = \Gamma^a(A)$ (by our remark concerning (P_0)).

The first question, we want to deal with is the question which varieties V have decidable first-order theories. There are only few examples known, e.g.

ABELIAN GROUPS (Szmielew 1949)

BOOLEAN ALGEBRAS (Tarski 1949)

RELATIVELY COMPLEMENTED DISTRIBUTIVE LATTICES (Ershov 1964)

BOOLEAN ALGEBRAS WITH A DISTINGUISHED ACTION OF A FINITE GROUP G AS AUTOMORPHISMS (Wolf - Burris 1975 - 1980)

COUNTABLE BOOLEAN ALGEBRAS WITH QUANTIFICATION OVER FILTERS (Rabin 1969)
(this of course is a second-order theory)

The first two results have been obtained by elimination of quantifiers, a very tedious method, but later Ershov invented the method of semantic embedding, i.e. he interpreted the language of relatively complemented distributive lattices into the language of Boolean algebras in such a way that a sentence is true for rel.compl. distr.lattices iff its translation is true for Boolean algebras.

This effective translation procedure together with Tarski's decision procedure for Boolean algebras now gives a decision procedure for rel. compl. distr. lattices. A short look at Foster's original definition of Boolean powers shows an immediate translation procedure for Boolean powers of A into the language of Boolean algebras by interpreting each variable as an A-sequence of variables in the language of Boolean algebras and then proceeding by induction on the complexity of the sentence and of the terms in the sentence. Thus the method of semantic embedding immediately applies to Boolean powers. There are two methods to push the result further to either Boolean products satisfying the maximum-property or else to sub-Boolean powers.

The first extension we get by consulting the Feferman-Vaught theorem, which gives an effective translation of the language of direct products of algebras into a sequence of sentences in the language of the given algebras and of Boolean algebras. This Feferman-Vaught-translation can also be used for Boolean products provided the maximum property holds and we get a decision procedure for the Boolean products provided we have it for the theory of stalks.

A has a decidable theory $\Rightarrow$ $\Gamma^e(A)$ has a decidable theory

For the second generalization we assume that all stalks in question are subalgebras of a finite algebra A . In this case the Ershov-translation as suggested by Foster's approach does not give a sentence in the language of A alone, but we also have to talk about the sets $X_P = \{x \in X \mid A_x \subseteq P\}$ which are closed subsets of X in the case of a sub-Boolean power and hence correspond to filters on the Boolean algebra. Thus we get a translation into the second order language of Boolean algebra with quantification over filters. Now Rabin's result gives a decision procedure for all countable sub-Boolean powers of A . A tedious construction shows, that for a finite set of finite algebras there is a finite algebra A , such that each countable member of $\Gamma^a(A)$ is isomorphic to a sub-Boolean power of A .

Consequently we have a second decidability result:

> If $\mathbb{A}$ is a finite set of finite algebras, then $\Gamma^a(\mathbb{A})$ has a decidable first-order theory.

Applying these results to discriminator varieties we obtain

<u>THM. 1</u>: If V is a discriminator variety generated by a finite algebra, then V has a decidable first-order theory.

<u>THM. 2</u>: Let $\mathbb{A}$ be the class of subdirectly irreducibles of a discriminator variety V, then V has a decidable first-order theory provided the theory $Th(\mathbb{A})$ of $\mathbb{A}$ is model-complete.

The proof of THM. 2 is based on the fact that each sentence is equivalent to some existential sentence modulo $Th(\mathbb{A})$ and hence the solution sets are precisely the solution sets of existential sentences which are always open. So each sentence has a clopen solution set and we have the maximum property, which shows $V = \Gamma^a(\mathbb{A}) = \Gamma^e(\mathbb{A})$.

S.Burris and R.McKenzie have extended this theory to congruence modular varieties using commutator theory and double-Boolean powers; similar techniques also lead to some scattered results on decidability and undecidability of first order theories of infinitely generated discriminator varieties.

Apart from decidability problems also several different model theoretic problems can be settled using Boolean constructions, e.g. the question, whether a given variety V has a model-companion. This problem can be rephrased into the question, whether the existentially closed members of V can be (first-order-)axiomatized. Also here the main tool is some translation-technique, in fact it is a small portion of the Feferman-Vaught-translation. Since an atomic sentence holds in a Boolean product iff its solution set is the whole base-space, the same is true for each existential conjunct of atomic formulas. This kind of analysis can be pushed further to all existential sentences. Having this translation it is possible to characterize the existentially closed members of the most interesting discriminator varieties.

THM. 3: If the class $\mathcal{A}$ of subdirectly irreducibles of a discriminator variety $\mathcal{V}$ has a model-companion $\mathcal{A}_o \subseteq \mathcal{A}$, then so does $\mathcal{V}$ itself and an axiom system for the model companion of $\mathcal{V}$ can be effectively constructed from an axiom system for $\mathcal{A}_o$.

As a byproduct the same techniques yield characterizations of special objects in a discriminator variety such as algebraically closed algebras, equationally compact algebras, topologically compact algebras, injective and weakly injective algebras. Several authors such as S.Burris, B.A.Davey, and P.Krauss have extended the results to other varieties than discriminator varieties using the following technique: first one has to prove that in the variety in question the Boolean products of simple members form an $\forall\exists$-axiomatic class and secondly one has to show that all the important special algebras in the variety are in fact such Boolean products.

A third model-theoretic problem that has been attacked by Boolean constructions is the search for $\aleph_o$-categorical algebras. $\aleph_o$-categorical algebras have been characterized in the variety generated by some finite field by Macintyre & Rosenstein and in the variety generated by a infra-primal algebra by D.M.Clark. The key-result is that a sub-Boolean power of a finite algebra A is $\aleph_o$-categorical iff the subalgebra generated by the closed sets X_P (P subalgebra of A) in the Heyting algebra of all closed sets is finite. Also this result is proved by translation techniques, but here the translation has to go both ways. One has to interpret the theory of the given variety in Boolean algebras and conversely the theory of Boolean algebras in the given variety.

Finally a further model theoretic problem was also attacked using Boolean powers, namely the construction of many nonisomorphic models in a given class of algebras. Early work on these lines was done by Tarski and Jónsson and the basic notion in this context is that of a B-separating algebra A, i.e. an algebra such that $A[B] \simeq A[C]$ implies $B \simeq C$. Now the fact that the variety of Boolean algebras has many nonisomorphic models in each cardinality carries over into the variety generated by A. Also many other product phenomena as listed in Hanf's paper carry over to varieties having a B-separating algebra. The technique of proving that an algebra is B-separating is again some translation-procedure. Either the algebra A contains a definable 2-element Boolean algebra, or it is simple and none of its finite powers has a skew congruence. In both cases the Boolean algebra B of A[B] is recognizable either from A[B] or its congruence lattice and hence $A[B] \simeq A[C]$ implies $B \simeq C$. A vast extension of this approach can be found in [12] in this volume.

The aim of this short survey was to show the reader an example of an algebraic construction which was not motivated from model theory but nevertheless is useful for solving model-theoretic problems. We also wanted to give the reader the flavor of the techniques used here without going into too many details, which can be very painful at some times. The reader interested in the details will have to consult the literature as listed in [2], [5], [14] . An excellent digest of the existing literature on Boolean powers is the paper [3].

REFERENCES

[1] R.F.AHRENS, I.KAPLANSKY : Topological representations of algebras. Trans. Amer. Math. Soc. 63 (1948), 457-481

[2] S.BURRIS : Boolean Powers, Alg. Univ. 5(1975),341-360

[3] S.BURRIS : Boolean Constructions. Proc.Cons. Puebla (1983), 67-90. Springer L.N. 1004

[4] S.BURRIS, H.P.SANKAPANAVAR : A course in universal algebra. Springer, New York (1981)

[5] S.BURRIS, R.McKENZIE : Decidability and Boolean representations. Memoirs Amer. Math. Soc. 246 (1981)

[6] D.CLARK, P.H.KRAUSS : Global subdirect products. Memoirs Amer. Math. Soc. 210 (1979)

[7] D.CLARK : $\aleph_0$-categoricity in infra-primal varieties. Algebra Univ. (in print)

[8] S.D.COMER : Elementary properties of structures of sections. Bol.Soc.Math. Mexicana 19 (1974), 78-85

[9] B.A.DAVEY, H.WERNER : Injectivity and Boolean powers. Math. Z. 166 (1979), 205-223

[10] A.L.FOSTER : Generalized "Boolean" theory of universal algebras I, II. Math. Z. 58 (1953), 306-336 / 59 (1953), 191-199

[11] W.HANF : On some fundamental problems concerning isomorphisms

[12] W.HODGES : On Constructing many non-isomorphic algebras. This volume.

[13] B.JÓNSSON : On isomorphism types of groups and other algebraic systems. Math. Scand 5 (1957), 224-229

[14] B.JÓNSSON : Review of Foster's 1961-paper, Math. Rev. 23, No. A. 85.

[15] H.WERNER : DISCRIMINATOR-Algebras. Akademie Verlag, Berlin (1978)

[16] H.WERNER : Sheaf-constructions and their elementary properties. Proc. Stefan Banach Center (1983)

Universal Algebra and its Links with Logic,
Algebra, Combinatorics, and Computer Science
Proc. "25. Arbeitstagung über Allgemeine Algebra", Darmstadt 1983
P. Burmeister et al. (eds.)

115 – 126

EMBEDDING PROBLEMS FOR RINGS AND SEMIGROUPS

P. M. Cohn

I. Uses of the Diamond Lemma

1. In many algebraic situations the elements of a set are defined as equivalence classes of formal expressions, where two expressions are considered equivalent if one can pass from one to the other by a series of 'moves'. The problem is to decide when two given expressions are equivalent.

The moves are usually of two sorts: direct moves (e.g. in a group, inserting xx^{-1}) and their inverses (removing a factor xx^{-1}). An expression is *reduced* if it admits no inverse moves, and the diamond lemma provides a simple hypothesis for each equivalence class (of the equivalence generated by all the moves) to contain just one reduced expression. Let us write $u \to v$ to indicate that we can pass from u to v by a direct move, and $u \twoheadrightarrow v$ to mean that we can pass from u to v by a finite sequence (possibly none) of direct moves. Inverse moves will be indicated by $\leftarrow$.

Diamond Lemma (Newman [12]). Let A be a set with an equivalence defined by moves as above, such that

i) (Finiteness condition) No element of A admits an infinite succession of inverse moves,

ii) (confluence condition) If $u \leftarrow v_1$ and $u \leftarrow v_2$ then there exists $w \in A$ such that $v_1 \twoheadleftarrow w$ and $v_2 \twoheadleftarrow w$.

Then each equivalence class of A contains exactly one reduced element.

The proof is straightforward: by (i) one has a reduced element in each equivalence class, and if there are two such elements joined by a sequence of (direct or inverse) moves, one modifies this sequence step by step in accordance with (ii).

There are a number of variants of this result (cf. Newman [32] also Pedersen [33]), but e.g. it is *not* enough to assume (drawing direct moves up and inverse ones down) that every descending path has a lower bound, cf. Newman [32].

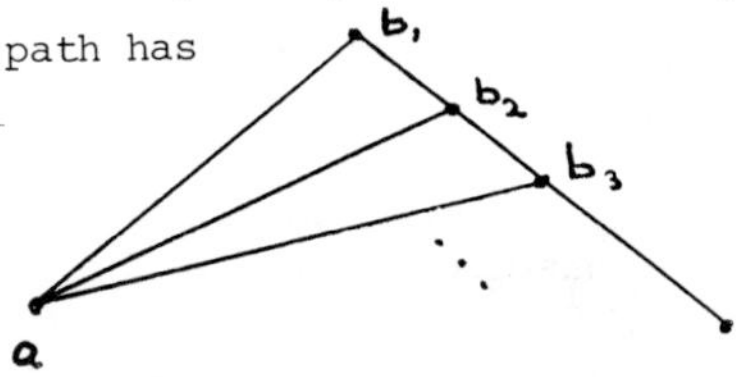

2. We now come to applications. α) The normal form for elements in a free group is

a good example. Let F be the free group on $a_1,\dots,a_r$ and consider words

$$w = a_{i_1}^{\varepsilon_1}\dots a_{i_n}^{\varepsilon_n} \qquad (\varepsilon_\nu = \pm 1).$$

Direct moves: $uv \to ulv$, $1 \to xx^{-1}$, where u,v are any expressions and x is a_i or a_i^{-1}. The inverse moves are just the inverses.

It is clear that (i) holds; we have to check (ii) and this is routine. The only case is essentially $xx^{-1}x \leftarrow 1.x$, $x.1$ and both of these $\leftarrow x$.

β) Here is a less trivial example. Let S be a semigroup without idempotent element. When can S be embedded in a semigroup without idempotent such that

Dr. $\qquad xa = b$ has a solution for all a,b?

We shall call this a Dr-*semigroup* (i.e. with right division). As a first guess one might expect left cancellation to be necessary, but in fact less is needed. A necessary condition for embeddability is

(1) $$ua = ub \Rightarrow va = vb \qquad \forall a,b,u,v \in S.$$

It turns out that this condition is also sufficient. The method of proof is to pick a,b ∈ S, adjoin formally a solution of

(2) $$xa = b,$$

and try to show that the resulting semigroup again satisfies (1). Calling the solution of (2) σ, we consider the set of all words *)

$$w = u_0\sigma^{\alpha_1}u_1\sigma^{\alpha_2}\dots\sigma^{\alpha_k}u_k,$$

where $u_i \in S$ for $1 \leqslant i \leqslant k-1$ and $u_0,u_k \in S \cup \{1\}$. We allow the direct move $b \to \sigma a$ with inverse $\sigma a \leftarrow b$, and use the diamond lemma to show that the new semigroup S' (with σ adjoined) satisfies (1) and that the natural mapping $S \to S'$ is injective (cf. [9]). Once this is established, a transfinite induction gives the desired Dr-semigroup (for another proof, using Croisot-Theissier semigroups, see Clifford-Preston [8] p.92).

γ) Our third example is an elaboration of β) and has an application to axiomatics of the real numbers. F.A. Behrend [1] proposed a set of axioms for ℝ involving semi-divisibility and asked (in a seminar in 1955) whether this implies commutativity. More precisely, he asked whether every semigroup satisfying Dr and

SDℓ. $\qquad$ For all a,b there exists z such that $a = b$ or $a = bz$ or $b = az$,

is necessarily a group. At first sight it is not obvious how to construct a

*) Strictly speaking, an equivalence on the set of these words has to be considered, see [9] for details.

counter-example. However, it is clear that such a semigroup must satisfy left-cancellation:

$C\ell$. For all a,b,u $ua = ub \Rightarrow a = b$.

If right cancellation also holds, we must have a group, so any counter-example will definitely not be a cancellation semigroup. One possible mode of construction is to take any semigroup with $C\ell$ and $SD\ell$ but without idempotent and as under β) adjoin a solution of an equation (2) (for a given pair of elements a,b) and repeat this for all possible pairs and show that the resulting semigroup S' again satisfies $C\ell$ and $SD\ell$ and has no idempotent, and the mapping $S \to S'$ is an embedding. This is once more an application of the diamond lemma [10].

δ) There is a corresponding but simpler result for groupoids, which states that any groupoid with left cancellation can be embedded in a groupoid with left divisibility (Evans [22]); this can be proved by using the semigroup presentation of groupoids and again invoking the diamond lemma (cf.[17], p.279).

ε) A typical application to ring theory is the proof of the Birkhoff-Witt theorem, particularly in the form used by Birkhoff [3]. This is so well known that we shall not weary the reader with the details, but just recall that from a basis (u_i) of a Lie algebra, taken totally ordered for convenience, we obtain a basis of the universal associative envelope U(L) in the form of all the ascending monomials $u_{i_1}u_{i_2}\dots u_{i_r}$ $(i_1 < i_2 < \dots < i_r)$. The 'straightening' process shows that we have a spanning set; to prove the linear independence we apply the diamond lemma, using the Jacobi-identity to verify the confluence condition.

3. There are many instances of the lemma being applied to rings (e.g. [11]), and this has led Bergman [2] to examine the special problems arising here and to give a general formulation of the result. One difficulty to be overcome is the following: Suppose that $u \leftarrow u^*$ is an inverse move, then we have the infinite chain

$$0 = u - u \leftarrow u^* - u \leftarrow u^* - u^* = 0 = u - u \leftarrow \dots$$

Of course, in any concrete case it is clear what has to be done to avoid this situation, but rather more care is needed in a general analysis.

Let k be a commutative ring (with $1 \neq 0$), X a set, <X> the free monoid on X and k<X> the free k-algebra on X (the monoid algebra on <X>). Let S be a subset of <X> × k<X>; its elements are written $\sigma = (W_\sigma, f_\sigma)$. For $A,B \in$ <X>

and $\sigma \in S$ we define the reduction $r_{A\sigma B}$ as the k-module endomorphism of $k\langle X\rangle$ which replaces $AW_\sigma B$ by $Af_\sigma B$ and leaves other elements of $\langle X\rangle$ unchanged. If no term $AW_\sigma B$ occurs in a, the reduction is *trivial* on a (it leaves a unchanged). If all reductions from S are trivial on a, a is *reduced*; if in any infinite chain of successive reductions on a only finitely many produce a change, a is *reduction-finite*. If all the reduced values obtained by reducing a are the same, a is called *reduction-unique*.

The diamond lemma for rings deals with two kinds of ambiguity: If A,B,C are monomials, then in reducing ABC,

i) An *overlap* ambiguity occurs when both AB, BC have the form W_{σ_i} (i=1,2) so that ABC can be reduced either to $f_{\sigma_1}C$ or Af_{σ_2}.

ii) An *inclusion* ambiguity results when both B, ABC have the form W_{σ_i}.

As a matter of fact, inclusion ambiguities are always avoidable; if B and ABC admit reductions, we can leave out the second. More precisely, we omit all reductions (W,f), where W contains the first component of another reduction as subword, and allow only *one* reduction with a given W as first component, but make appropriate allowance for the omitted reduction.

An ambiguity is *resolvable* if the two expressions obtained can be reduced to the same element of $k\langle X\rangle$.

Theorem (Bergman [2], see also [17] p.341). Let S be a reduction system for $k\langle X\rangle$ *such that all elements are reduction-finite. Then the elements of* $k\langle X\rangle$ *are all reduction-unique if and only if all ambiguities of S are resolvable.*

Of course this is merely a restatement of the diamond lemma in a form suitable for use in rings, and we shall say no more about the proof, but add a few general remarks on the way it is used.

The moves effecting the reduction arise from a particular presentation of our rings as quotient of $k\langle X\rangle$. When choosing the reduction system we first take the relations given by the presentation. Each ambiguity leads to a further relation and some skill and luck is needed to find a system satisfying the above hypotheses. *Skill*, because it may happen for certain presentations that some choices of reduction system do not lead to a terminating reduction procedure (more and more ambiguities appear), and *luck*, because for some presentations there may be no reduction system with a terminating procedure. For a finitely presented algebra may well have an insoluble word problem. This just means that our sets of relations are not recursive, but trivially they are always recursively enumerable and the above theorem just describes ways of enumerating them or rather, the (possibly infinite) reduction system. We refer to Bergman [2] for a further discussion of these points and some illuminating examples.

II. Localization in non-commutative rings

1. I want to consider a problem for rings and semigroups which from its statement certainly seems to belong to universal algebra. Although it is a normal form problem, the diamond lemma is of no help. I am speaking of the following embedding problems:

ER. When can a ring R be embedded in a skew field?

ES. When can a semigroup S be embedded in a group?

It is clear that in ES we may take S to have a 1, i.e. we need only deal with monoids. ES was solved in 1939 by Mal'cev [29], who gave as necessary and sufficient condition an infinite sequence of conditions of the form

$$A_1 \wedge \ldots \wedge A_r \Rightarrow A,$$

where the A's are atomic formulae. Such conditions are called quasi-identities (in a quasi-identity A may also be f = falsity); this was to be expected because the classes defined by quasi-identities, *quasi-varieties* are just universal classes admitting direct products (cf. e.g. [17] p.235), and the class of monoids embeddable in groups evidently forms a quasi-variety.

In 1971 I solved ER (cf. [12],[13] Ch.7), again by giving an infinite sequence of conditions [16]. Of course this time we cannot expect just quasi-identities, because the class of rings embeddable in skew fields does not admit products and so is not a quasi-variety, but it comes very close to being one. Let us define an *integral domain* as a non-zero ring without zero-divisors and denote the class of integral domains by $\mathcal{D}_0$, and the class of rings embeddable in skew fields by $\mathcal{E}$. Then we have

Theorem 1. There is a quasi-variety $\mathcal{T}$ *such that* $\mathcal{E} = \mathcal{T} \cap \mathcal{D}_0$.

This tells us that the class $\mathcal{E}$ is characterized by quasi-identities plus

$$\forall x,y \quad xy = 0 \Rightarrow x = 0 \vee y = 0; \qquad 1 \neq 0.$$

The proof in outline goes as follows. A ring R is called *strongly regular* if for each $a \in R$ there exists $x \in R$ such that $a^2x = a$. Now $\mathcal{T}$ is the class of rings embeddable in strongly regular rings. Clearly $\mathcal{T}$ is universal and admits direct products, hence it is a quasi-variety. Further, any skew field clearly is strongly regular, so $\mathcal{E} \subseteq \mathcal{T} \cap \mathcal{D}_0$. To establish equality here one proves first that every strongly regular ring is a subdirect product of skew fields, and then shows that an integral domain which is embedded in a direct product of skew fields is itself embeddable in a skew field (cf. [14]).

2. Let us now look at the embedding problem ER in detail. It is interesting to observe that although both in the semigroup case and the ring case the actual quasi-identities are quite complicated, they can be much better organized for rings. As an illustration we note that Klein [27] was able, by writing Mal'cev's semigroup conditions in matrix form, to prove

Theorem 2. If an integral domain R has the property that any n×n nilpotent matrix A over R satisfies $A^n = 0$, *then* $R^{\cdot}$ *is embeddable in a group.*

Here $R^{\cdot}$ denotes the set of non-zero elements of R. Klein also observes that the condition in his theorem is not necessary for $R^{\cdot}$ to be embeddable in a group.

Let R be any ring and S a subset of R, then a homomorphism $f: R \to R'$ is called *S-inverting* if the elements of S map to invertible elements of R'. The *universal S-inverting ring* R_S is a ring with an S-inverting homomorphism $\lambda: R \to R_S$ which is universal with this property, i.e. the diagram shown can always be completed in just one way. It is easy to see that R_S exists: we need only adjoin formal inverses of the elements of S to R. The problem is that there may be no convenient way of expressing the elements of R_S, to say nothing of a normal form (which need not exist even in the commutative case, cf. [18]).

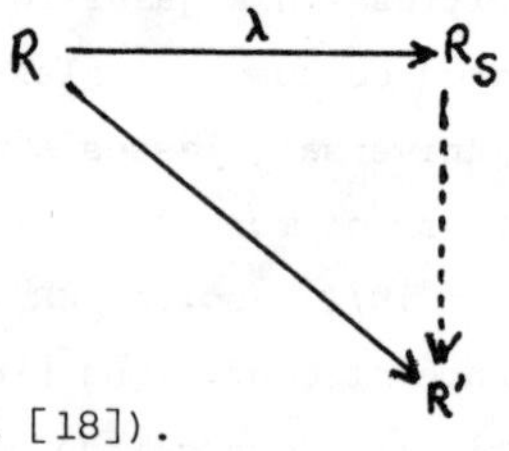

We digress for a moment to consider some related problems (cf. Bokut' [6]). Let us again write $\mathcal{D}_0$ for the class of integral domains, $\mathcal{D}_1$ for the class of rings R such that $R^{\cdot}$ is embeddable in a group, and $\mathcal{D}_2$ for the class of rings R such that the universal $R^{\cdot}$-inverting map is injective, then it is clear that

$$\mathcal{D}_0 \supseteq \mathcal{D}_1 \supseteq \mathcal{D}_2 \supseteq \mathcal{E}.$$

In fact all these inequalities are strict: $\mathcal{D}_0 \neq \mathcal{D}_1$ was proved by Mal'cev [28] in 1937 in his solution of v.d. Waerden's problem (whether $\mathcal{D}_0 = \mathcal{E}$), $\mathcal{D}_1 \neq \mathcal{D}_2$ was proved by Bokut' [4,5] in 1967 in solving Mal'cev's problem (whether $\mathcal{D}_1 = \mathcal{E}$), while $\mathcal{D}_2 \neq \mathcal{E}$ follows from the work of Bowtell [7], Klein [26] and also from recent results of Gerasimov [23].

To return to our problem: the ring R_S is generated by $R \cup S^{-1}$, but in a complicated way, which makes it hard to tell when an expression represents 0. To remedy this defect, we generalize the problem by considering matrices rather than elements. Thus let Σ be a set of square matrices over R and define R_Σ as before as the universal Σ-inverting ring. It may be constructed by introducing for each matrix $A = (a_{ij}) \in \Sigma$ another matrix $A' = (a'_{ij})$ with defining relations (in matrix form): $AA' = A'A = I$. In the commutative case one needs to assume S to be multiplicative to get a convenient expression for the elements of R_S. Similarly we shall here need to assume Σ *multiplicative*; this means that $1 \in \Sigma$ and if $A,B \in \Sigma$, then $\begin{pmatrix} A & C \\ O & B \end{pmatrix} \in \Sigma$ for every C of appropriate size. With this definition we have

Theorem 3. Let R be any ring and Σ a multiplicative set of square matrices over R. Then the elements of the universal Σ-inverting ring R_Σ can all be obtained as entries of inverses of matrices in Σ:

$$R_\Sigma = \{a'_{ij} \mid (a'_{ij}) = A^{-1},\ A \in \Sigma\}.$$

This is quite straightforward to prove ([13] p.250); an equivalent formulation is:

Theorem 3'. Let R,Σ be as in Th.3, then the entries of R_Σ can be obtained as components of the solutions of equations

$$Ax = b \qquad \text{where } A \in \Sigma \text{ and } b \text{ is a column over } R.$$

3. So much is formal, but the actual problem ER remains. It may be generalized by asking

K. What is the kernel of the canonical mapping $\lambda: R \to R_\Sigma$?

As an intermediate step between ER and K one can study the homomorphisms from R to fields (we shall henceforth omit the prefix 'skew'). A field K with a homomorphism $f: R \to K$ is called an R-*field*; if K is generated by the image of f, we speak of an *epic* R-field (this is precisely an epimorphism in the category of rings). To study the homomorphism $f: R \to K$, define Ker f, the *singular kernel* of f, as the set of all square matrices over R which are mapped to singular matrices over K by f. One can write down conditions for a set of matrices to be a singular kernel, which are reminiscent of the conditions for a prime ideal in a commutative ring, and for this reason these sets of matrices are also called *prime matrix ideals*. Thus one obtains a bijection between the set of prime matrix ideals of R and epic R-fields, which enables one to solve ER and more generally describe the epic R-fields in terms of R alone. This was done in Ch.7 of [13]; I shall not repeat the details here but consider the problem K more closely.

Here the theory of prime matrix ideals is of no help. We recall that if $\mathcal{P}$ is a prime matrix ideal of R and Σ its complement in the set of all square matrices over R, then R_Σ is a local ring and its residue class field $K_{\mathcal{P}}$ with canonical homomorphism $\phi: R \to R_\Sigma \to K_{\mathcal{P}}$ is such that $\text{Ker}\phi = \mathcal{P}$. But this method tells us nothing about the kernel of $\lambda: R \to R_\Sigma$; in fact, since R_Σ is not usually a field, we cannot speak of a singular kernel here.

4. In order to study the kernel of $\lambda: R \to R_\Sigma$ one has to modify the above approach. This is recent work by Gerasimov [24] and Malcolmson [30,31], and I want to describe the main idea briefly.

We start again from R_Σ and take $A \in \Sigma$ but instead of looking at the entries of A^{-1} (Th.3) or $A^{-1}b$ (Th.3') we consider the element $cA^{-1}b$, where $c \in R^n$, $A \in \Sigma_n$ ($n\times n$ matrices in Σ), $b \in {}^nR$. It is suggestive to write this as an $(n+1)\times(n+1)$ matrix:

$$\begin{pmatrix} A & b \\ c & 0 \end{pmatrix} \quad \text{or, on changing notation,} \quad a = \begin{pmatrix} A & 'a \\ a' & 0 \end{pmatrix}. \tag{3}$$

Denote the set of all such matrices (for $n = 1,2,\ldots$) by $\mathfrak{M}$. We can construct R_Σ directly from $\mathfrak{M}$ as follows. We define operations $\oplus$, $\odot$, $\ominus$ by

$$a \oplus b = \begin{pmatrix} A & 0 & 'a \\ 0 & B & 'b \\ a' & b' & 0 \end{pmatrix} \quad a \odot b = \begin{pmatrix} A & -'ab' & 0 \\ 0 & B & 'b \\ a' & 0 & 0 \end{pmatrix} \quad \ominus a = \begin{pmatrix} A & -'a \\ a' & 0 \end{pmatrix}.$$

We shall admit the following elementary transformations on matrices of $\mathfrak{M}$:

E.1. To a given row (column) block add a left multiple of a later row (right multiple of an earlier column) block.

This just amounts to left (right) multiplication by an upper unitriangular matrix.

A square block T on the main diagonal of A in (3) is called *superfluous* if either the row block or the column block containing T is 0, apart from T itself. Now we have

E.2. Insert (or remove) a superfluous block and the row and column block in which it lies.

We write $a \sim b$ (for $a,b \in \mathfrak{M}$) if we can pass from a to b by a sequence of elementary transformations E.1,2. Clearly this defines an equivalence relation on $\mathfrak{M}$; the class of a will be written as [a]. With these definitions we have

Theorem 4. $\mathfrak{M}/\sim$ *is the universal Σ-inverting ring R_Σ; the canonical homomorphism $\lambda: R \to R_\Sigma$ is given by*

$$a \longrightarrow \left[\begin{pmatrix} 1 & -a \\ 1 & 0 \end{pmatrix}\right].$$

The proof is fairly routine. As an example consider the identity $a\oplus(\ominus a) = 0$. This is proved by the following reduction, where dots stand for zeros:

$$\begin{pmatrix} A & \cdot & 'a \\ \cdot & A & -'a \\ a' & a' & \cdot \end{pmatrix} \to \begin{pmatrix} A & A & \cdot \\ \cdot & A & -'a \\ a' & a' & \cdot \end{pmatrix} \to \begin{pmatrix} A & \cdot & \cdot \\ \cdot & A & -'a \\ a' & \cdot & \cdot \end{pmatrix} \to \begin{pmatrix} A & \cdot \\ a' & \cdot \end{pmatrix} \to (\cdot) .$$

The interest of this result is that it allows an explicit determination of the kernel $\lambda: R \to R_\Sigma$. We have

Malcolmson's criterion [30]. *Let* $a = \begin{pmatrix} A & 'a \\ a' & \cdot \end{pmatrix}$, *then* $[a] = 0$ *in* R_Σ *if and only if*

$$\begin{pmatrix} A & \cdot & \cdot & 'a \\ \cdot & F & \cdot & \cdot \\ \cdot & \cdot & G & 'g \\ a' & f' & \cdot & \cdot \end{pmatrix} = \begin{pmatrix} P \\ u \end{pmatrix} (Q \quad v)$$

for $F, G, P, Q \in \bar{\Sigma}$ *and a suitable row* u *and column* v.

Here $\bar{\Sigma}$ is the set of all block-triangular matrices with blocks on the main diagonal from $\Sigma \cup \{1\}$, zeros below and arbitrary matrices above.

The criterion is not easy to apply in practice, but then that was not to be expected. Nevertheless Gerasimov and Malcolmson are able to prove the following results (recall that an n-fir is a ring over which every n-generator right ideal is free, of unique rank):

Theorem 5 [24]. *If* R *is an n-fir* $(n \geq 2)$, *then* $R_{R^\cdot}$ *is an* $(n-2)$*-fir*,

Theorem 6 [32]. *If* R *is an n-fir* $(n \geq 2k)$, *and* Σ *the set of all* $k\times k$ *matrices with no zero-divisor as factor, then* R_Σ *is an* $(n-2k)$*-fir.*

Moreover, these methods will also give the construction of an epic R-field from a prime matrix ideal. To indicate this briefly, we first note a consequence of Th.4, which uses the matrix analogue of ideals (cf. [13] p.260).

Corollary to Th.4. Let R_Σ *be as in Th.4 and* $\mathfrak{A}$ *a matrix ideal in* R *such that* $A \dot{+} B \in \mathfrak{A}$, $A \in \bar{\Sigma}$ *implies* $B \in \mathfrak{A}$. *Then* $\mathfrak{A}$ *consists of full* $\sim$*-classes in* $\mathfrak{M}$ *and* $[\mathfrak{A}] = \{[a] \mid a \in \mathfrak{A}\}$ *is an ideal in* R_Σ.

Now it is not hard to prove that if Σ is the complement of a prime matrix ideal $\mathfrak{P}$, then R_Σ is a local ring (cf.[15]). Let $\mathfrak{P}_0$ be the singular kernel of the mapping $R \to R_\Sigma \to K_{\mathfrak{P}}$; since the mapping is Σ-inverting, any matrix not in $\mathfrak{P}$ is inverted, so we have $\mathfrak{P}_0 \subseteq \mathfrak{P}$. By the above Corollary, $[\mathfrak{P}]$ is an ideal in R_Σ which is clearly proper, but $[\mathfrak{P}_0]$ is the unique maximal ideal, hence $[\mathfrak{P}_0] = [\mathfrak{P}]$ and it follows that $\mathfrak{P}_0 = \mathfrak{P}$.

5. We recall that a *semifir* is a ring which is an n-fir for all n and a *full* matrix over a semifir is a square matrix with no zero-divisor as factor (cf. [13] Ch.7). It is not hard to verify that over a semifir the non-full matrices form the unique least prime matrix ideal. This means that a semifir has a universal field of fractions, inverting all full matrices. More generally, this is true for the Sylvester domains (defined by Sylvester's law of nullity, cf. Dicks-Sontag [21]); in fact this is the precise class of rings R with an R-field inverting all full matrices. A slight generalization is possible by using instead the matrices that are *stably full*, i.e. such that $A \dotplus I_n$ is full for all n (Pseudo-Sylvester domains, cf. [20]). More generally one could consider all matrices A such that $A \dotplus \dots \dotplus A \dotplus I_n$ is full, but this raises new problems (the corresponding rank function is no longer integer-valued). But it directs attention to rank functions on rings, which are far from completely explored (cf. Goodearl-Handelman [25], Schofield [34], also [19]).

References

1. F.A. Behrend, A system of independent axioms for magnitudes, J. and Proc. Roy. Soc. New South Wales 37 (1953) 27-30.

2. G.M. Bergman, The diamond lemma for ring theory, Advances in Math. 29 (1978) 178-218.

3. G. Birkhoff, Representability of Lie algebras and Lie groups by matrices, Ann. of Math. 38 (1937) 526-532.

4. L.A. Bokut, On the embedding of rings in skew fields (Russian) DAN SSSR 175 (1967) 755-758.

5. L.A. Bokut, On Mal'cev's problem (Russian) Sibirsk. Mat. Zhurnal 10 (1969) 965-1005.

6. L.A. Bokut, Associative Rings 2 (Russian) Novosibirsk NGU, 1981.

7. A.J. Bowtell, On a question of Mal'cev, J. Algebra 7 (1967) 126-139.

8. A.H. Clifford and G.B. Preston, The algebraic theory of semigroups, vol.II, Amer. Math. Soc., Providence R.I., 1967.

9. P.M. Cohn, Embeddings in semigroups with one-sided division, J. London Math. Soc. 31 (1956) 169-181.

10. P.M. Cohn, Embeddings in sesquilateral division semigroups, J. London Math. Soc. 31 (1956) 181-191.

11. P.M. Cohn, Some remarks on the invariant basis property, Topology 5 (1966) 215-228.

12. P.M. Cohn, Un critère d'immersibilité d'un anneau dans un corps gauche, C.R. Acad. Sci. Paris, Ser.A, 272 (1971) 1442-1444

13. P.M. Cohn, Free rings and their relations, Academic Press, London, New York 1971.

14. P.M. Cohn, Rings of fractions, Amer. Math. Monthly 78 (1971) 596-615.

15. P.M. Cohn, Universal skew fields of fractions, Symposia Math. VIII (1972) 135-148.

16. P.M. Cohn, The class of rings embeddable in skew fields, Bull. London Math. Soc. 6 (1974) 147-148.

17. P.M. Cohn, Universal Algebra 2nd ed. Reidel, Dordrecht 1981.

18. P.M. Cohn, The universal field of fractions of a semifir I. Numerators and denominators, Proc. London Math. Soc. (3) 44 (1982) 1-32.

19. P.M. Cohn, Rank functions on rings, to appear.

20. P.M. Cohn and A.H. Schofield, On the law of nullity, Math. Proc. Camb. Phil. Soc. 91 (1982) 357-374.

21. W. Dicks and E.D. Sontag, Sylvester domains, J. Pure and Applied Algebra 13 (1978) 243-275.

22. T. Evans, On multiplicative systems defined by generators and relations I. Normal form theorems, Proc. Camb. Phil. Soc. 47 (1951) 637-649.

23. V.N. Gerasimov, Inverting homomorphisms of rings (Russian), Algebra i Logika 18, No.6 (1979) 648-663.

24. V.N. Gerasimov, Localization in associative rings (Russian), Sibirsk. Mat. Zh. 23 (1982) 36-54.

25. K.R. Goodearl and D. Handelman, Rank functions and K_0 of regular rings, J. Pure and Applied Algebra 7 (1976) 195-216.

26. A.A. Klein, Rings nonembeddable in fields with multiplicative semigroups embeddable in groups, J. Algebra 7 (1967) 100-125.

27. A.A. Klein, Necessary conditions for embedding rings into fields, Trans. Amer. Math. Soc. 137 (1969) 141-151.

28. A.I. Mal'cev, On the immersion of an algebraic ring into a field, Math. Ann. 113 (1937) 686-691.

29. A.I. Mal'cev, Über die Einbettung von assoziativen Systemen in Gruppen (Russian, German summary) I. Mat. Sbornik 6 (48) (1939) 331-336, II. Mat. Sbornik 8 (50) (1940) 251-264.

30. P. Malcolmson, Construction of universal matrix localizations, in Advances in non-commutative ring theory (ed. P.J. Fleury), Lecture Notes in

Math. vol. 951, Springer Berlin 1982, pp.117-131.

31. P. Malcolmson, Matrix localizations of n-firs, Trans. Amer. Math. Soc. (in press).

32. M.H.A. Newman, On theories with a combinatorial definition of "equivalence", Ann. of Math. 43 (1942) 223-243.

33. J. Pedersen, Some confluence results, to appear.

34. A.H. Schofield, Simple artinian rings (1983 Adams Prize Essay) to appear.

Bedford College,
Regent's Park
London NW1 4NS

Universal Algebra and its Links with Logic,
Algebra, Combinatorics, and Computer Science
Proc. "25. Arbeitstagung über Allgemeine Algebra", Darmstadt 1983
P. Burmeister et al. (eds.)

127 – 142

VARIETÄTEN VON MONOIDEN, KONGRUENZEN UND SPRACHEN – ODER: WIE MAN ZÄHLT

H. Jürgensen[1)]

Als M.P. Schützenberger vor ungefähr 20 Jahren eine kombinatorische Charakterisierung der Monoide mit nur trivialen Untergruppen veröffentlichte [Sch1], war in ihrem Gefolge und zusätzlich durch zahlreiche Fragen der Codierungstheorie motiviert die Frage zu stellen, ob es insgesamt gelingen könnte, algebraisch vernünftige Klassen von Halbgruppen kombinatorisch vernünftig zu charakterisieren.

In dem vorliegenden Beitrag werden einige Gedankengänge zu diesem Thema verfolgt und einige typische Problemkreise angesprochen. Anspruch auf Vollständigkeit erheben wir nicht; das Literaturverzeichnis erwähnt aus der Fülle des Relevanten nur das direkt Benutzte. Viel Wichtiges wurde der Umfangsbeschränkung geopfert.

Als Werkzeug, Zusammenhänge zwischen algebraischen und kombinatorischen Eigenschaften von Halbgruppen herzustellen, dienen die sogenannten Hauptkongruenzen: S sei eine Halbgruppe[2)], $S_1 \subseteq S$,

---

1) Department of Computer Science, The University of Western Ontario, London, Ont., Canada N6A 5B7

2) Wie üblich, ist $S^1 = S$, falls S ein Monoid ist, und $= S \cup \{1\}$ mit $1s=s1=s$ für $s \in S^1$, falls S kein Monoid ist.

$S_2 \subseteq S^1$, $L \subseteq S_1 \cap S_2$. Die Mengen S_1, S_2 und L definieren folgendermaßen eine Äquivalenzrelation auf S_1:

$$\forall y,z \in S_1 \left(y \equiv z \left(P_L^{(S_1,S_2)}\right) \Leftrightarrow \left(\forall u,v \in S_2\colon uyv \in L \leftrightarrow uzv \in L\right)\right)$$

Übliche Spezialfälle sind:

(1) $S_1 = S$, $S_2 = S$. Dann ist $P_L^{(S_1,S_2)}$ eine Kongruenzrelation auf S, die *Hauptkongruenz* P_L von L im Sinne von Dubreil [Du1].

(2) $S_1 = S$, $S_2 = S^1$. Dann ist $P_L^{(S_1,S_2)}$ ebenfalls eine Kongruenzrelation auf S, die *syntaktische Kongruenz* P_L^1 von L. Die Relation P_L^1 ist die gröbste Kongruenz auf S, welche L saturiert. Man nennt

$$\text{Syn } L = S/P_L^1$$

die *syntaktische Halbgruppe* von L. L heißt *disjunktiv* in S, falls P_L^1 die Gleichheitsrelation auf S ist. Folglich ist L/P_L^1 immer in Syn L disjunktiv.

Ist X eine Menge, so bezeichnet X* die Menge aller Wörter über X; X* ist ein Monoid mit dem leeren Wort 1 als Einselement.

Im folgenden werden wir fast ausschließlich P_L^1 betrachten. Wir beginnen mit einigen Beispielen:

Beispiel 1. G sei eine Gruppe, N ein echter Normalteiler von G. $A \subseteq G$ sei ein Repräsentantensystem der Restklassen von G/N. Sei $\bar{L} \subseteq A$ und $L = \bigcup_{g \in \bar{L}} gN$. Dann ist $P_L^1 = P_L$ eine Kongruenz auf G, und zwar gehört zu P_L ein Normalteiler N_L mit $N \subseteq N_L \subseteq G$. Folglich ist

$$\text{Syn } L = G/N_L = (G/N) \;/\; (N_L/N)$$

Faktorgruppe von G/N.

Beispiel 2. Sei $X = \{a\}$, $L = \{1,a\} \cup \{a^{2n} \mid n \geq 1\} \subseteq X^*$, $S = X^*$. Dann gibt es die folgenden P_L^1-Klassen - es ist wieder $P_L^1 = P_L$:

$$\{1\},\ \{a\},\ \{a^{2n} \mid n \geq 1\},\ \{a^{2n+1} \mid n \geq 1\}.$$

Die Multiplikation von Syn L ist durch die folgende Tafel gegeben:

	1	a	a^2	a^3
1	1	a	a^2	a^3
a	a	a^2	a^3	a^2
a^2	a^2	a^3	a^2	a^3
a^3	a^3	a^2	a^3	a^2

Die Greenschen Relationen in Syn L sind:

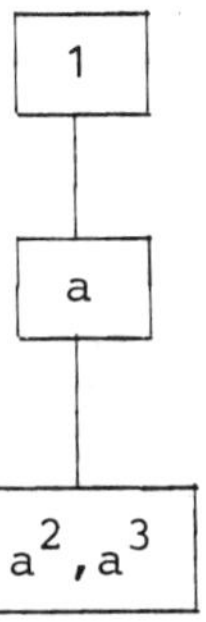

In diesem Falle hat man $\mathcal{D} = \mathcal{H}$, und Syn L hat die zwei Untergruppen $\{1\}$ und $\{a^2,a^3\} \simeq Z_2$. Die rechtsreguläre Darstellung von Syn L hat den Graphen:

$$1 \xrightarrow{a} a \xrightarrow{a} a^2 \overset{a}{\longleftrightarrow} a^3$$

Der nicht-trivialen Untergruppe $\{a^2,a^3\}$ entspricht der Zykel im Graphen und der * in der regulären Darstellung 1+a+(aa)* von L. Im Graphen von Syn L ist L durch diejenige Menge von Pfaden charakterisiert, die von 1 zu einem der Elemente von L/P_L führen; in diesem Falle ist $L/P_L = \{1,a,a^2\}$.

<u>Beispiel 3</u>. In dem Monoid $S = \{1,r_1,r_2,r_3\}$ mit der Multiplikation

$$1x = x1 = x \text{ für } x \in S$$

und

$$r_i r_j = r_j \text{ für } i,j = 1,2,3$$

ist keine Teilmenge disjunktiv. Es gibt also keine Halbgruppe $\bar{S}$ und keine Menge $L \subseteq \bar{S}$, so daß $\bar{S}/P_L^1 \simeq S$ ist.

Beispiel 4. Sei $X = \{[,]\}$ und $L \subseteq X^*$ die folgendermaßen induktiv definierte Menge:

(1) $1 \in L$,

(2) $u \in L \to [u] \in L$,

(3) $u, v \in L \to uv \in L$.

(4) Nur Elemente gemäß 1-3 sind in L.

Die Menge L ist eine *Dyck-Sprache*. Syn L ist dem bizyklischen Monoid

$$B = \langle\, a,b \mid ab = 1 \,\rangle$$

isomorph. Der Isomorphismus wird von der Abbildung

$$X = \{[,]\} \to \{a,b\}: \begin{cases} [\mapsto a \\ ] \mapsto b \end{cases}$$

induziert. Die zu $b^i a^j$ gehörige P_L^1-Klasse ist die Klasse der $u \in X^*$ mit *Defekt* (i,j). Dabei ist der Defekt $\delta(u)$ eines Wortes $u \in X^*$ folgendermaßen definiert:

$$\delta(u) = \begin{cases} (0,0), & \text{falls } u = 1, \\ \delta(v) + (0,1), & \text{falls } u = v[, \\ \delta(v) - (0,1), & \text{falls } u = v] \text{ und } \delta(v)_2 > 0, \\ \delta(v) + (1,0), & \text{falls } u = v] \text{ und } \delta(v)_2 = 0. \end{cases}$$

Man beachte, daß im Beispiel 4

$$L = 1/P_L^1 \quad \text{und Syn } L = \langle\, X \mid u = 1 \text{ für } u \in L \,\rangle$$

gilt. Die syntaktische Kongruenz P_L^1 und die durch die Präsentation mit Relationensystem

$$\{u = 1 \mid u \in L\}$$

definierte Kongruenz ρ_L stimmen überein. Im allgemeinen sind P_L^1

und ρ_L jedoch Antipoden, wie das folgende, vorerst letzte Beispiel zeigt.

Beispiel 5. Man betrachte das Monoid $M = \{1, z_2, z_3\}$ mit der durch

$$1x = x1 = x \text{ für } x \in M$$

und

$$z_i z_j = z_i \text{ für } i,j = 1,2$$

gegebenen Multiplikation. Es sei $\nu \in \text{Endo } M$ der triviale Endomorphismus mit $\nu(M) = \{1\}$. S sei die Bruck-Reilly-Erweiterung [Jü1]

$$BR(M,\nu) = N_O \times M \times N_O.$$

Sei $L = \{(0,1,0)\}$. Es gilt

$$(n,z_1,m) \equiv (n,z_2,m) \quad (P_L^1)$$

für alle n und alle m. Folglich ist $S/P_L^1 \not\simeq S$. Sei nun X ein Erzeugendensystem von S, und sei φ der durch die Inklusion $X \subseteq S$ induzierte surjektive Homomorphismus von X* auf S, und sei $\bar{L} = \varphi^{-1}(L)$. Dann ist S homomorphes Bild von $X^*/\rho_{\bar{L}}$ mit

$$\rho_{\bar{L}} = \{u = 1 \mid u \in \bar{L}\},$$

und $X^*/P_{\bar{L}}^1 \simeq S/P_L^1$ wird folglich nicht durch $\langle X \mid \rho_{\bar{L}} \rangle$ präsentiert.

Diese Beispiele - wie viele andere - zeigen, daß offenbar ein gewisser Zusammenhang zwischen kombinatorischen Eigenschaften von Mengen und algebraischen Eigenschaften von Monoiden besteht, daß er aber doch sehr kompliziert werden kann und sich insbesondere einer einfachen Behandlung durch Präsentationen entzieht.

Kommen wir auf die eingangs erwähnte Charakterisierung der Monoide mit nur trivialen Untergruppen zurück! Um diese und weitere Aussagen formulieren zu können, definieren wir zunächst, was wir unter *zählen* verstehen wollen. Es sei X eine endliche Menge, $L \subseteq X^*$

und $w \in X^*$. Dann sei

$$w_L = \{u \mid u \in X^*X \cap L,\ w \in uX^*\}.$$

Für $m \in N$, $r \in N_0$ mit $0 \leq r < m$ sei

$$(L,r,m) = \{w \mid w \in X^*,\ |w_L| \equiv r \pmod m\}.$$

Dabei ist N die Menge der natürlichen Zahlen und $N_0 = N \cup \{0\}$. Intuitiv gesprochen, ist (L,r,m) die Menge derjenigen Wörter über X, die man mit einem mod-m-Zähler durch L herausfiltern kann Dadurch wird es möglich, Wortmengen zu charakterisieren, deren syntaktische Monoide nur p-Gruppen oder nur auflösbare Gruppen enthalten.

Definition 6. Zu jeder endlichen Menge X sei *Aper* X die kleinste Teilmenge F von 2^{X^*} mit den folgenden Eigenschaften:

(1) $\forall a \in X: \{a\} \in F$

(2) $L, L' \in F \rightarrow \left(L \cup L' \in F \wedge X^* \setminus L \in F \right)$

(3) $L, L' \in F \rightarrow LL' \in F$

Satz 7 (Schützenberger, Straubing[1]). Sei X eine endliche Menge und $L \subseteq X^*$. Dann gilt $L \in$ *Aper* X genau dann, wenn Syn L endlich ist und nur triviale Untergruppen besitzt.

Die Aussage von Satz 7 wie auch die Operationen von Definition 6 sind gewissermaßen charakteristisch für das mit endlichen syntaktischen Monoiden Erreichbare.

Definition 8. Unter einer *Pseudovarietät von Monoiden* versteht man eine Klasse *M* von Monoiden, die die folgenden Abschlußeigenschaf-

---

1) Vgl. [St1].

ten hat[1]:

(1) $M_1 \in \mathcal{M}$, $M_2 \leq M_1 \rightarrow M_2 \in \mathcal{M}$

(2) $M_1, M_2 \in \mathcal{M} \rightarrow M_1 \times M_2 \in \mathcal{M}$

(3) $M_1 \in \mathcal{M}$, φ ein Homomorphismus von $M_1 \rightarrow \varphi M_1 \in \mathcal{M}$

Die Abweichung von der üblichen Varietätendefinition wird gemacht, um insbesondere Klassen von ausschließlich endlichen Monoiden erfassen zu können. Der Schritt über endliche Monoide hinaus ist bei der vorliegenden Fragestellung problematisch.

Es sei $\Xi = \{x_1, x_2, \ldots\}$ eine abzählbare Menge von Symbolen und

$F = \{u_i = v_i \mid i = 1,2,\ldots\}$

eine Folge von Gleichungen mit $u_i, v_i \in \Xi^*$. Ein Monoid M *erfüllt die Gleichung* $u_i = v_i$, wenn für jeden Homomorphismus $\varphi : \Xi^* \rightarrow M$ auch $\varphi u_i = \varphi v_i$ gilt. M *erfüllt* F *schließlich*, wenn M fast alle Gleichungen von F erfüllt.

Es sei $\mathcal{V}_F$ die Klasse aller endlichen Monoide, die F schließlich erfüllen, also

$$\mathcal{V}_F = \bigcup_{k=1}^{\infty} \bigcap_{i=k}^{\infty} \mathcal{V}_{(u_i = v_i)},$$

wobei $\mathcal{V}_{(u_i=v_i)}$ die Klasse aller endlichen Monoide ist, die $u_i = v_i$ erfüllen. Man zeigt, daß $\mathcal{V}_{(u_i=v_i)}$ und $\mathcal{V}_F$ Pseudovarietäten von Monoiden sind [Ei1]. $\mathcal{V}_F$ ist die *durch* F *schließlich definierte Pseudovarietät von endlichen Monoiden*. Eine Pseudovarietät $\mathcal{V}$ von Monoiden ist *schließlich gleichungsdefiniert*, wenn es eine Folge F von Gleichungen gibt, so daß $\mathcal{V}_F = \mathcal{V}$ gilt.

---

1) Wir schreiben $\leq$ für die Untermonoidrelation.

Satz 9 (Eilenberg). Jede nicht-leere Pseudovarietät von endlichen Monoiden ist schließlich gleichungsdefiniert.

Beispiel 10.

(a) $F = \{x^n = x^{n+1} \mid n \geq 1\}$ definiert schließlich die Pseudovarietät der endlichen Monoide mit nur trivialen Untergruppen

(b) Für $n \in N$ sei $\bar{n} = \mathrm{kgV}(1,2,\ldots,n)$. Die Folge $F = \{x^{\bar{n}} = 1 \mid n \geq 1\}$ definiert schließlich die Pseudovarietät der endlichen Gruppen

Bevor wir das kombinatorische Äquivalent betrachten, wollen wir zwei Analoga des Satzes 7 angeben, die zur Motivation der weiteren Überlegungen beitragen können.

Satz 11 (Straubing). Zu jeder endlichen Menge X sei *MAufl* X die kleinste Teilmenge F von 2^{X^*} mit den folgenden Eigenschaften:

(1) $\forall a \in X\colon \{a\} \in F$

(2) $L, L' \in F \rightarrow \left(L \cup L' \in F \wedge X^* \setminus L \in F \right)$

(3) $L, L' \in F \rightarrow LL' \in F$

(4) $L \in F \rightarrow \forall a \in X\ \forall m \in N, m \geq 2\ \forall 0 \leq r < m\colon (La, r, m) \in F$

Es gilt $L \in$ *MAufl* X genau dann, wenn Syn L endlich ist und nur auflösbare Untergruppen besitzt.

Satz 12 (Kleene). Zu jeder endlichen Menge X sei *Rat* X die kleinste Teilmenge F von 2^{X^*} mit den folgenden Eigenschaften:

(1) $\forall a \in X\colon \{a\} \in F$

(2) $L, L' \in F \rightarrow \left(L \cup L' \in F \wedge X^* \setminus L \in F \right)$

(3) $L, L' \in F \rightarrow LL' \in F$

(4) $L \in F \rightarrow L^* \in F$

Es gilt $L \in$ *Rat* X genau dann, wenn Syn L endlich ist.

Es ist an der Zeit, die in den letzten Sätzen betrachtete Situation zu verallgemeinern:

Definition 13. Eine *Mengen-*-Varietät* V ordnet jeder endlichen Menge X eine Menge $VX \subseteq 2^{X^*}$ zu, so daß die folgenden Bedingungen erfüllt sind:

(1) $\forall X\ \forall L,L' \in VX$: $L \cup L' \in VX \wedge X^* \setminus L \in VX$

(2l) $\forall X\ \forall L \in VX\ \forall a \in X$: $a^{[-1]}L \in VX$

(2r) $\forall X\ \forall L \in VX\ \forall a \in X$: $La^{[-1]} \in VX$

(3) $\forall X,Y\ \forall L \in VX\ \forall \varphi \in Hom(Y^*,X^*)$: $\varphi^{-1}L \in VY$

Dabei ist

$$a^{[-1]}L = \{w \mid aw \in L\},\ La^{[-1]} = \{w \mid wa \in L\}.$$

Satz 14 (Eilenberg). M sei eine Pseudovarietät endlicher Monoide. Dann ist LM mit

$$LMX = \{L \mid L \subseteq X^*,\ \mathrm{Syn}\ L \in M\}$$

eine Varietät von rationalen Sprachen[1]. Ist L eine Varietät von rationalen Sprachen, so sei ML die von der Klasse

$$\{\mathrm{Syn}\ L \mid L \in LX \text{ für ein } X\}$$

erzeugte Pseudovarietät endlicher Monoide. Dann gilt

$$MLM = M \text{ und } LML = L.$$

Es soll noch auf eine weitere zum bisherigen Kontext gehörige Begriffsbildung eingegangen werden, die von Thérien [Th1] vorgeschlagen wurde.

---

1) "Sprachenvarietät" hier gleich "Mengen-*-Varietät". Es ist $LMX \subseteq \mathcal{R}at X$.

Definition 15 (Thérien). Eine *Kongruenzen-*-Varietät* ordnet jeder endlichen Menge X eine Menge KX von Kongruenzen auf X^* zu, so daß die folgenden Bedingungen erfüllt sind:

(1) $\forall X\ \forall \alpha_1,\alpha_2 \in KX$: $\alpha_1 \cap \alpha_2 \in KX$

(2) $\forall X,Y,M\ \forall \alpha \in KX\ \forall \varphi \in Hom(Y^*,X^*)\ \forall \psi \in Hom(X^*/\alpha, M)$:

Die durch $\varphi\bar{\alpha}\psi$ definierte Kongruenzrelation liegt in KY. Dabei ist M ein beliebiges Monoid und $\bar{\alpha}$ der kanonische Homomorphismus von X^* auf X^*/α

Beispiel 16.

(1) $KX = \{\alpha \mid \exists n \in N\ \forall x \in X^*: x^n \alpha x^{n+1}\}$ gehört zu *Aper* X.

(2) $KX = \{\alpha \mid \exists n \in N\ \forall x \in X^*: x^n \alpha 1\}$ definiert die endlichen Gruppen.

Satz 17 (Thérien). K sei eine *-Varietät von Kongruenzen von endlichem Index. Dann ist LK mit

$$LKX = \{L \mid L \subseteq X^*,\ P_L^1 \in KX\}$$

eine Varietät von rationalen Sprachen. Ist L eine Varietät von rationalen Sprachen, so sei KL die von den Mengen

$$KLX = \{P_L^1 \mid L \in LX\}$$

erzeugte Kongruenzen-*-Varietät. Dann gilt

$$KLK = K \text{ und } LKL = L.$$

Wir wenden uns jetzt erneut einer Definition des Zählens zu: Sei $m \in N$, $t \in N_0$. Für $a,b \in N_0$ gelte

$$a\ \vartheta_{t,m}\ b \Leftrightarrow \begin{cases} a<t \text{ und } b=a \\ \text{oder} \\ a \geq t,\ b \geq t \text{ und } a \equiv b \pmod{m}. \end{cases}$$

$\vartheta_{t,m}$ zählt mit Schwellwert t und modulo m:

$$0 \longrightarrow 1 \longrightarrow 2 \ \ldots \longrightarrow t-1 \longrightarrow t \longrightarrow t+1 \ \ldots \longrightarrow t+m-1$$

Sei nun X eine nicht-leere endliche Menge. Für $u,v \in X^*$ sei

$$\binom{u}{v}$$

die Anzahl der Möglichkeiten, u als verstreutes Teilwort in v einzubetten[1]. Für $l \in N_0$ und $v,w \in X^*$ gelte

$$v\ \alpha_{t,m,l}\ w \Leftrightarrow \forall u \in X^* \setminus X^* X^{l+1}: \binom{v}{u}\ \vartheta_{t,m}\ \binom{w}{u}.$$

$\alpha_{t,m,l}$ ist eine Kongruenz von endlichem Index. Sei nun β eine beliebige Kongruenz auf X^* und $u,v \in X^*$ mit $u=u_1\ldots u_n$, $u_1,\ldots,u_n \in X$; sei $Z = (z_0,\ldots,z_n)$ eine Folge von Wörtern $z_0,\ldots,z_n \in X^*$; wir definieren

$$\binom{v}{u}_{z\beta}$$

als die Anzahl der Möglichkeiten, u als verstreutes Teilwort in v in der Form

$$v = x_0 u_1 x_1 u_2 x_2 \ldots u_n x_n$$

mit

$$x_i\ \beta\ z_i \text{ für } i = 0,1,\ldots,n$$

einzubetten. Sei dann

$$v\ (\beta\alpha_{t,m,l})\ w \Leftrightarrow \forall u \in X^* \setminus X^* X^{l+1}\ \forall z: \binom{u}{v}_{z\beta}\ \vartheta_{t,m}\ \binom{w}{u}_{z\beta}.$$

Man definiert jetzt eine Folge von Kongruenzen:

$$\alpha^0_{t,m,l} = \omega = X^* \times X^*,$$

$$\alpha^{i+1}_{t,m,l} = (\alpha^i_{t,m,l}\ \alpha_{t,m,l}).$$

Es sei nun

$$K^i_{t,m,l}X = \{\alpha \mid \alpha \text{ ist Kongruenz auf } X^* \text{ mit } \alpha^i_{t,m,l} \le \alpha\}.$$

Ferner seien Schreibweisen wie

$$K^i_{t,*,l}X = \bigcup_{m \ge 0} K^i_{t,m,l}X$$

vereinbart.

---

1) Zu den verallgemeinerten Binomialkoeffizienten vergleiche [Ei1] und [Oc1].

<u>Satz 18</u> (Thérien). Jedes $K^{i}_{t,m,l}$, $K^{i}_{*,m,l}$ usw. ist eine Kongruenzen-*-Varietät. Insbesondere gelten die folgenden Korrespondenzen:

$K^{1}_{0,*,*}$: nilpotente Gruppen

$K^{1}_{0,*,1}$: abelsche Gruppen

$K^{*}_{0,*,*}$: auflösbare Gruppen

$K^{*}_{*,1,*}$: Monoide mit nur trivialen Untergruppen

$K^{*}_{*,*,*}$: Monoide mit nur auflösbaren Untergruppen

Damit ist Zählen, soweit es das durch auflösbare Gruppen Beschreibbare betrifft, in einem in jeder Hinsicht natürlichen Detail beschrieben. Neben vielem anderen ist im Bereich der rationalen Sprachen allerdings die Frage, wie man in nicht-auflösbaren Gruppen "zählt", immer noch offen.

Geht man über den Bereich der rationalen Sprachen und damit der endlichen Monoide hinaus, so trifft man eine wesentlich kompliziertere Situation an. Satz 14 gilt in dieser Form nicht mehr (vgl. u.a. [Pe1]). Daß das syntaktische Monoid auch einer noch recht einfachen Sprache ebenfalls für recht komplizierte Sprachen syntaktisch sein kann, zeigt das Beispiel der Dycksprache und auch das der in X* disjunktiven Sprachen [Sh1].

Nicht jede natürliche Klasse rationaler Sprachen, die eine brauchbare Charakterisierung durch syntaktische Halbgruppen[1)] oder Monoide besitzt, ist eine Varietät (für Beispiele vgl. [Va1]). Auch speziel-le, gegenüber zusätzlichen Operationen abgeschlossene Varietäten wur-

1) In gewissen Fällen ist es günstiger, Sprachen in $X^{+}=XX^{*}$, syntaktische Halbgruppen und, dazu gehörig, +-Varietäten zu betrachten.

den untersucht (z.B. [St2]).

Was von dem Genannten überträgt sich, wenn man statt endlicher Folgen, d.h. statt der Wörter, unendliche Folgen über einem endlichen Alphabet X betrachtet?

Es sei X^{ω} die Menge der (einseitig) unendlichen Folgen über X, der *ω-Wörter*, und es sei $X^{\infty} = X^{\omega} \cup X^*$. Zu $L \subseteq X^{\infty}$ gehört die in naheliegender Weise verallgemeinerte syntaktische Kongruenz P_L^1 auf X^* und das syntaktische Monoid Syn $L = X^*/P_L^1$ [Jü4+5].

Satz 19 ([Jü3]). Jedes endlich erzeugte Monoid ist dem syntaktischen Monoid einer Menge $L \subseteq X^{\infty}$ isomorph.

Es ist dabei interessant festzustellen, daß zur Konstruktion einer Menge $L \subseteq X^{\infty}$ mit vorgegebenem syntaktischem Monoid M im allgemeinen als Alphabet ein minimales Erzeugendensystem von M nicht ausreicht. Wann zusätzliche Symbole - maximal 2 - erforderlich sind und welches ihre genaue Funktion ist, ist ein offenes Problem; Beispiele legen den Eindruck nahe, daß sie den Takt für das Zählen erzeugen - man bedenke, daß gegenüber den endlichen Wörtern bei den ω-Wörtern ein Zählbezugspunkt, nämlich das Wortende, fehlt.

Unter den Teilmengen von X^{∞} spielen die mit trivialem syntaktischen Monoid eine wichtige Sonderrolle - hierin weicht X^{∞} wesentlich von X^* ab. Eine Menge $L \subseteq X^{\infty}$ heißt *absolut abgeschlossen*, wenn $X^*L \subseteq L$ und Suff $L \subseteq L$ gilt; dabei ist Suff L die Menge der Suffixe von L.

Satz 20 ([Jü6]). Eine Menge $L \subseteq X^{\infty}$ ist genau dann absolut abgeschlossen, wenn Syn L trivial ist.

Zu $L \subseteq X^{\infty}$ sei L_{cl} die maximale in L enthaltene absolut abgeschlossene Menge, und es sei

$$\mathrm{Red}\, L = L \setminus L_{cl}.$$

Die Menge

$$\lambda L = \{L' \mid L' \subseteq X^{\infty},\ L' = \bar{L} \cup \mathrm{Red} L \text{ für } \bar{L} \text{ absolut abgeschlossen, } \bar{L} \cap \mathrm{Red} L = \emptyset\}$$

bildet einen Verband mit 0 und 1. Die Restriktion $\lambda_* L$ von λL auf X^* ist im allgemeinen trivial. Genauer gilt

$$\lambda_* L = \{L \cap X^*\}$$

für $L \cap X^* \neq \emptyset, X^*$ und

$$\lambda_* L = \{\emptyset, X^*\}$$

für $L \cap X^* \in \{\emptyset, X^*\}$. Dagegen enthält die Restriktion $\lambda_{\omega} L$ von λL auf X^{ω} so interessante Beispiele wie die Menge aller disjunktiven ω-Wörter.

Im Zusammenhang mit Varietäten von Teilmengen von X^{∞} spielt λL eine wichtige Rolle.

Satz 21 ([Jü3]). M sei eine Klasse von Monoiden. Zu jedem Alphabet X sei $L_{\infty} M X$ die Klasse der Mengen $L \subseteq X^{\infty}$ mit Syn $L \in M$. Ist M eine Pseudovarietät, so erfüllt $L_{\infty} M$ die Bedingungen 1,21 und 3 aus Definition 13, und für $L \in L_{\infty} M X$ gilt $\lambda L \subseteq L_{\infty} M X$.

Literaturhinweise

[Br1] J.A. Brzozowski: Developments in the Theory of Regular Languages. In: S.H. Lavington (Hrsg.), Information Processing 80, 29-40. North Holland Publ. Co., Amsterdam, 1980.

[Cl1] A.H. Clifford, G.B. Preston: The Algebraic Theory of Semigroups. I, II. American Mathematical Society, Providence, Rh.I., 1961, 1967.

[Du1] P. Dubreil: Contribution à la théorie des demi-groupes. Mém. Acad. Sci. Inst. France (2), 63 (1941).

[Ei1] S. Eilenberg: Automata. Languages, Machines. A, B. Academic Press, New York, 1974, 1976.

[Jü1] H. Jürgensen: Total disjunktive verallgemeinerte Bruck-Reilly-Erweiterungen von Halbgruppen und formale Sprachen. In: K.H. Hofmann, H. Jürgensen, H.J. Weinert (Hrsg.), Recent Developments in the Algebraic, Analytical, and Topological Theory of Semigroups. Lecture Notes in Math. 998, 281-309. Springer Verlag, Berlin, 1983.

[Jü2] H. Jürgensen, H.J. Shyr, G. Thierrin: Monoids with Disjunctive Identity and Codes.

[Jü3] H. Jürgensen, G. Thierrin: Which Monoids Are Syntactic Monoids of ω-Languages?

[Jü4] H. Jürgensen, H.J. Shyr, G. Thierrin: Properties of the ω-Languages in Relation with Some of Their Associated Equivalences. Bericht TI 7/80, Institut für Theoretische Informatik, TH Darmstadt, 1980.

[Jü5] H. Jürgensen, H.J. Shyr, G. Thierrin: Disjunctive ω-Languages. EIK 19 (1983), 267-278.

[Jü6] H. Jürgensen, G. Thierrin: On ω-Languages Whose Syntactic Monoid is Trivial. International J. of Computer and Information Sci. 12 (1983), 359-365.

[Kl1] S.C. Kleene: Representation of Events in Nerve Nets and Finite Automata. Shannon, McCarthy (Hrsg.), Automata studies. Princeton Univ. Press, 1954. 3-41.

[Oc1] P. Ochsenschläger: Binomialkoeffizienten in freien Monoiden. Dissertation, Darmstadt, 1981.

[Pe1] J.-F. Perrot, J. Sakarovitch: A Theory of Syntactic Monoids for Context Free Languages. In: B. Gilchrist (Hrsg.), Information Processing 1977, 69-72. North-Holland Publ. Co., Amsterdam, 1977.

[Pe2] J.-F. Perrot: Variétés de langages et opérations. Theoret. Comp. Sci. 7 (1978), 197-210.

[Pi1] J.-E. Pin: Sur le monoide syntactique de L* lorsque L est un langage fini. Theoret. Comp. Sci. 7 (1978), 211-215.

[Sa1] J. Sakarovitch: Monoides pointés. Semigroup Forum 18 (1979), 235-264.

[Sch1] M.P. Schützenberger: On Finite Monoids Having only Trivial Subgroups. Information and Control 8 (1965), 190-194.

[Sh1] H.J. Shyr: Disjunctive Languages on a Free Monoid. Information and Control 34 (1977), 123-129.

[St1] H. Straubing: Families of Recognizable Sets Corresponding to Certain Varieties of Finite Monoids. J. Pure and Appl. Algebra 15 (1979), 305-318.

[St2] H. Straubing: Aperiodic Homomorphisms and the Concatenation Product of Recognizable Sets. J. Pure and Applied Algebra 15 (1979), 319ff.

[Th1] D. Thérien: Classification of Regular Languages by Congruences. Report CS-80-19. University of Waterloo, Dep. of Computer Science.

[Va1] E. Valkema: Zur Charakterisierung formaler Sprachen durch Halbgruppen. Dissertation, Kiel, 1974.

Universal Algebra and its Links with Logic,
Algebra, Combinatorics, and Computer Science
Proc. "25. Arbeitstagung über Allgemeine Algebra", Darmstadt 1983
P. Burmeister et al. (eds.)

143 – 149

DIAGRAMS, EQUATIONS AND THEORIES IN CATEGORIES

S. Mac Lane

Many sorted universal algebra manipulates equations - by symmetry, transitivity and especially by substitution. The study of algebraic objects internal to a category constructs and manipulates commutative diagrams. The first observation of this paper is a simple but useful metatheorem asserting that these two types of manipulations are essentially the same. Moreover this simple metatheorem can save all manner of minor troubles - for example, the theory of internal categories in an elementary topos.

The second observation deals with "essentially" algebraic theories as described by a "sketch" in the sense of Bastiani - Ehresmann, and summarizes a proof by G.M. Kelly of a theorem of those authors on the completeness of the category in which such a sketch may be embedded.

Finally, an third observation will involve some more general comments on the connections, both past and prospective, between universal algebra and related disciplines, such as category theory or other parts of algebra.

1. Equations and Diagrams. A semigroup is described (as usual) as a set S with a binary operation $m: S \times S \to S$ (of multiplication) which satisfies the associative law. Correspondingly one may define a semigroup object (an "internal" semigroup) S is any category with products as an object S of the category together with an arrow $m: S \times S \to S$ for which the usual diagram expressing the associative law is commutative. In much the same way one defines group objects, ring objects and module objects relative to a ring object in a category. Our general metatheorem applies to any such type of algebraic theory. Indeed, consider a many-sorted algebraic theory Ω with sorts $X_1, X_2, \ldots$ and primitive operations $m : X_1 \times \ldots \times X_k \to X'$ of various arities $k \geq 0$; these operations are subjected to axioms in the form of equations $s = t$ between terms s and t . Here the terms $t(x,y,\ldots)$ are constructed by composition from a stock of variables of each sort and from primitive terms

$m(x_1, x_2, \ldots, x_k)$. A model of the theory Ω then consists of sets $S_1, S_2, \ldots$, one of each sort, together with functions m_S which satisfy the equations. Here we may apply the usual rules for deriving new equations. These rules (written, for simplicity, with a single variable, are the following:

(i) Reflexivity, Symmetry and Transitivity of equality.

(ii) Substitution of an equation in a term: From an equation $s = t$ between two terms of type Z and a term $u(z)$ with a variable z of type Z, to derive the equation $u(s) = u(t)$.

(iii) Substitution in an equation: Given an equation $s(x) = t(x)$, to replace a variable x by any term v of the same type.

Now let C be a category with products. For each such type Ω of algebraic theory one then may describe an Ω-object in C to be a string of objects $X_1, X_2, \ldots$ of C , one of each sort, together with arrows m . The primitive equations are then represented by primitive communtative diagrams in C . Our metatheorem then asserts that any deduction of additional equations according to the standard rules (i) - (iii) listed above corresponds to the construction of new commutative diagrams from the given primitive ones, and conversely. In stating this metatheorem, each term $t(x,y)$ of type A with variables x and y of sorts X and Y is interpreted as an arrow $|t| : X \times Y \to A$ between the (products of) the corresponding objects, and each step (i), (ii) or (iii) corresponds to an "elementary" commutative diagram. One also uses elementary diagrams expressing the fact that the product $X \times Y$ in the category is a bifunctor; if $f: X \to X'$ and $g: Y \to Y'$, then $(f \times 1) \circ (1 \times g) = (1 \times g) \circ (f \times 1) : X \times Y \to X' \times Y'$. It is then straight-forward to prove that every deduction according to the rules corresponds to the construction of a commutative diagram in C , and conversely.

As equations are usually interpreted in sets the "variables" x range over elements of a set X . In the categorical interpretation, the objects X of the category C do not have any "elements", but the variables x are represented as identity arrows $x: X \to X$. This is a special case of the interpretation of logistic formulas in a category, as in the Mitchell-Benabou language (see Johnstone [2]).

This metatheorem saves considerable trouble in the construction of large commutative diagrams. For example if the category C has pullbacks, the metatheorem applies also to algebraic systmes with partially defined operations such as categories themselves. For example, if A_o is the object of objects and A_1 the object of arrows of a category object A with domain and codomain arrows $\partial_o, \partial_1 : A_1 \to A_o$, then the composition of arrows in the category A is defined to be an arrow from the pullback $A_1 \times_{A_o} A_1$ of the diagram

$$A_1 \xrightarrow{\partial_o} A_o \xleftarrow{\partial_1} A_1 .$$

In this way one can define in the ambient category $\mathbb{C}$ such things as functor objects $F: A \to B$ and adjoint functors. The metatheorem then tells us at once that the whole elementary theory of the properties of adjoint functors carries over from ordinary category theory (in sets) to category objects and functor objects in $\mathbb{C}$. This "carry over" includes the various definitions of an adjunction, by a bijection between hom-sets, by universal arrows or by unit and counit of the adjunction. The equivalence between these alternative definitions (as presented for example in Mac Lane [5]) thus applies at once to the general situation of categories in a topos. In the formulation it is convenient to use for the hom sets the language described by Johnstone and Wraith in [3].

This metatheorem thus makes very simple the otherwise complex description of internal category theory, as presented for example in Chapter 2 of Johnstone [2].

For classical universal algebra the k-ary operations are defined on products $X_1 \times X_2 \times \dots$. There is a more general siutation where the operations are defined on pullbacks (over prior operations), just as for the definition given above for composition in a category. In this case, following Freyd, one speaks of an "essentially" algebraic theory. It is very useful to extend universal algebra to such cases - which are considerably more specific than "partial" functions.

2. Sketches can be complete. It is well known that a many-sorted algebraic theory can be described as a category S with certain specified limits (for example, all finite products of the sorts), with suitable arrows (including the primitive operations of all arities); a model of such a theory can then be described as a product - preserving functor on such a category S to sets. This description includes in particular the Lawvere theories (one-sorted), as explained for example in Richter [6].

There have been many different extensions and generalizations of this description of an algebraic theory, so as to include for example essentially algebraic theories. One of the most flexible extensions is the notion of a "sketch", as formulated by Ehresmann. Such a sketch is a category with a certain list of specified "cones", which are intended to be limiting cones in the model (for example, to be the projections of a product or a pullback).

More formally, recall that a small category L is a category in which the class of objects and the class of arrows are both sets. All our definitions will be relative if a given set F of small categories L ; in particular, F might be the set of all finite categories - or perhaps just one copy of each isomorphism type of each such. If $G: L \to S$ is a functor, recall that a cone φ on G with base L consists of

(i) An object A of S ;

(ii) The constant functor $\Delta A: L \to S$;

(iii) A natural transformation $\varphi: \Delta A \to G$.

Such a natural transformation φ consists of a commutative diagram of arrows $\varphi_{\ell}: A \to G_{\ell}$, one for each object $\ell \in L$, all from the common domain A ; hence the term "cone". Such a cone φ is said to be a limiting cone if for each object B of S and each cone $\beta: \Delta B \to G$ there is a unique arrow $f: B \to A$ such that $\beta = \varphi \Delta_f$. For example, if L is that discrete category which is a set of just two objects ℓ and ℓ' then a limiting cone for G is just the product $G_{\ell} \times G_{\ell'}$ with its projections onto the factors $G\ell$ and $G\ell'$.

An F-sketch (S, Φ) is now defined to be a small category S together with a set Φ of cones on functors $G: L \to S$, where each base L is a category in F . If $\mathbf{C}$ is a category which is F-complete (i.e., which has limits for all F-based cones) then a model of the sketch (S, Φ) in $\mathbf{C}$ is a functor $H: S \to \mathbf{C}$ which takes each cone $\varphi \in \Phi$ onto a limiting cone. The category $\mathrm{mod}(S, \Phi ; \mathbf{C})$ of all such models is the collection of all "algebras" in $\mathbf{C}$ for the theory represented by the sketch S - and the basic idea is that each model must turn every cone φ in the sketch into a limiting cone in the model.

In the best cases one can get all the algebras by using only those sketches (T, Ψ) in which all the cones of Ψ are already limiting cones in T - for example, because T is itself already F-complete. This is already the case for Lawvere theories (which have in T all powers of the single underlying sort) or for many-sorted theories with all possible products (or pullbacks) of sorts. The cited theorem of Bastiani - Ehresmann asserts that this can always be accomplished by sufficiently "enlarging" the category S of a given sketch.

Theorem. To every F-sketch (S, Φ) there exists an F-complete small category T, an F-sketch (T, Ψ) in T and a model M of (S, Φ) in T which is generic in the following sense: For every F- complete category $\mathcal{C}$, composition with M induces on the models a functor

$$M^{*}: \mathrm{Mod}(T, \Psi ; \mathcal{C}) \longrightarrow \mathrm{Mod}(S, \Phi ; \mathcal{C}))$$

which is an equivalence of categories.

This means that each model $H: T \rightarrow \mathcal{C}$ of the new sketch Ψ yields by composition $H \mapsto H \circ M: S \rightarrow \mathcal{C}$ a model of the originally given sketch; and that this functor M^{*} is an equivalence in the sense that there is a functor N in the direction opposite that of M^{*} with both composites $M^{*}N$ and NM^{*} naturally isomorphic to the identity functor. In other words, the new sketch (T, Ψ) has essentially the same category of models as the originally given sketch (S, Φ).

The original proof [1] of the theorem by Bastiani and Ehresmann built up the requisite complete category T by suitable transfinite induction. The new proof by Kelly [4] is simpler and more conceptual. Moreover, it provides a clear conceptual description of the functor N "converse to M^{*}". With the generic model M one has for each model H of T a diagram

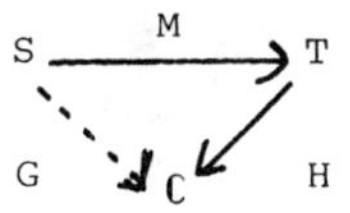

in which H determines G as the composite $G = H \circ M$. If we write $\mathcal{C}^{T}$ for the functor category of all such H, this is a functor

$$M^{*}: \mathcal{C}^{T} \longrightarrow \mathcal{C}^{S}$$

Now in such circumstances, this functor M^{*} may have a right adjoint, called the right Kan extension along M. It assigns to each G the "best" extension to T along M. Now Kelly's proof of the theorem shows that this (pointwise) right Kan extension exists and is the functor N which makes M^{*} the asserted equivalence between the categories of models.

This clear and explicit result indicates that the categorical approach provides an effective way of generalizing the notion of an algebraic theory.

3. Interconnections. These two items are examples of some of the interconnections between universal algebra and category theory; there are many more such. Because of them, it does seem to me that this conference might have done well to add to the various topics "Universal algebra and X " a topic "Universal algebra and category theory". It is unfortunately the case that at present both category theory and Universal algebra suffer too much from a growing isolation from other mathematical activities. In neither case was this the original intention.

Universal algebra began with the clear objective of finding and formulating those general theorems (such as the Jordan-Hölder theorem or the Krull-Schmidt theorem) which apply to many different kinds of algebraic systems. Similarly, lattice theory began with the intent of finding those properties of subsystems of algebraic systems which have a general import. Today in universal algebra and lattice theory the emphasis seems to be directed more toward the solution of hard but sometimes artificial problems, not necessarily related to the rest of algebra. There is moreover an unfortunate habit of comparing "general algebra" to "special algebra" - as if these were two equal halves of a whole. They are not.

There are similar difficulties in category theory. The original objective was the formulation of a few simple but general concepts which would help in the conceptual understanding of various topics in mathematics. In some directions, such as topos theory and the understanding of set-theoretic forcing, this objective is still in view. In some other parts of category theory, however, the variety of general concepts has now multiplied (many special sorts of categories and too many definitions of theories). As a result, the original objective seems almost lost in a maze of generality.

In both of these fields it would be in order now to emphasize the original objectives of the field, and to observe the extensive interconnections of each field with all the rest of algebra and indeed with the rest of mathematics.

The University of Chicago

References

[1] Bastiani, Andrée and Charles Ehresmann. Categories of sketched structures. Cahiers Topologie Geom Differentielle 13 (1972) 103-214.

[2] Johnstone, Peter T. Topos Theory. London Math. Soc. Mono No. 10, Academic Press, London, New York, San Francisco 1977.

[3] Johnstone, Peter T. and Gavin C. Wraith. Algebraic Theories in Toposes, pp 141-242 in Indexed categories and their applications, Springer lecture notes in Mathematics vol 661, Springer Verlag, Berlin 1979.

[4] Kelly, G.M. On the essentially algebraic theory generated by a sketch, Bull. Australian Math. Soc. 26 (1982) 45-56.

[5] Mac Lane, Saunders. Categories for the working mathematician, Graduate Texts in Mathematics No. 5. Springer Verlag, Berlin 1971.

[6] Richter, Gunther. Kategorielle Algebra. Akademie Verlag, Berlin 1979.

Universal Algebra and its Links with Logic,
Algebra, Combinatorics, and Computer Science
Proc. "25. Arbeitstagung über Allgemeine Algebra", Darmstadt 1983
P. Burmeister et al. (eds.)

150 – 167

UNIVERSAL ALGEBRA AND COMBINATORICS

R. W. Quackenbush

1. Introduction

The connections between universal algebra and combinatorics have been well exposed in recent years, especially by Trevor Evans (see [2], [3]). In this paper I want to discuss three areas of combinatorics where ideas from universal algebra are playing a significant role. The first is the result of Eric Mendelsohn that every finite group is isomorphic to the automorphism group of a finite Steiner triple system. The second is Bernhard Ganter's approach to coding theory via universal algebra. The third is the area of enumerative combinatorics, combining category theory, classical algebra and universal algebra. The use of category theory and classical algebra to do enumerative combinatorics is now well established, through the efforts of Gian-Carlo Rota, his students and colleagues. The addition of universal algebra as an enumerative tool is due to Anatol Meush.

2. Automorphism groups of finite Steiner triple systems.

Steiner triple systems are the most interesting and most widely studied class of block designs; $\langle P, B\rangle$ is a Steiner triple system (STS) if P is a set (of points) and B is a set of 3-element subsets of P (called blocks) such that for distinct p, $q \in P$ there is a unique $b \in B$ containing both p and q. A straightforward counting argument shows that if $|P|$ is finite, then $|P| \equiv 1$ or 3 mod 6. Conversely, but less straightforwardly, this necessary condition for the existence of a finite STS is sufficient.

There are two familiar classes of STSs: projective geometries over GF(2) and affine geometries over GF(3). We will be interested in the former. Since a vector space over GF(2) is equivalent to a boolean group (i.e., a group satisfying $a^2 = 1$), let us start with a boolean group $\langle P'; +, 0\rangle$ (with the group operation written additively). With $P = P'-\{0\}$, set

$$B = \{\{a,b,a+b\} \mid a,b \in P,\ a \neq b\}.$$

It is easy to check that $\langle P, B\rangle$ is an STS.

In checking that this construction works, we make use of certain identities true in boolean groups, but we do not make use of the full power of the associative law. The identities that we need are:

$$a+0 = a,\ a+a = 0,\ a+b = b+a,\ a+(a+b) = b.$$

An algebra $\langle A; +, 0\rangle$ satisfying these identities is called a Steiner loop (sloop); all sloops give rise to STSs. Conversely, if $\langle P, B\rangle$ is an STS, then $\langle A; +, 0\rangle$ is a sloop if we define A to be $P \dot{\cup} \{0\}$ and define + by $a+0 = 0+a = a$ and $a+a = 0$ for all $a \in A$ and for distinct $a,b \in P$, $a+b = c$ where $\{a,b,c\} \in B$.

Which groups are automorphism groups of STSs? This is easily answered: every group is isomorphic to the automorphism group of some STS. Given an arbitrary group $\underline{G}$, it is easy to construct a partial STS (e.g. where distinct points belong to at most one block) whose automorphism group is isomorphic to $\underline{G}$; if $\underline{G}$ is finite, then the partial STS will also be finite. Then one merely completes the partial STS freely to an STS. This construction has the drawback that the completed STS will be infinite even when the partial STS is finite. Since STSs are really only of interest in the finite case, this is a fatal drawback. What, then, are the automorphism groups of finite STSs?

If we try to complete a finite partial STS to a finite STS, we are very likely to find that the non-trivial automorphisms of the partial STS do not extend to automorphisms of the completed STS. Thus, a different approach is called for.

Given a finite group $\underline{G}$, it is easy to find a finite STS into whose automorphism group $\underline{G}$ can be embedded: if $|G| = n$, take $\langle P_n, B_n\rangle$ to be the STS corresponding to $\underline{2}^n$, the nth power of the 2-element group. As it is an n-generated free boolean group, the subgroup of its automorphism group obtained by permuting some free generating set is $\underline{S}_n$, the symmetric group on n letters. The idea now is to change B_n so as to kill off all automorphisms not corresponding to some fixed embedding of $\underline{G}$ into $\mathrm{Aut}(\langle P_n, B_n\rangle)$ and to preserve those which do correspond to $\underline{G}$ (at the same time we must insure that no new automorphisms arise).

Fortunately, it is quite easy to perturb the blocks of an STS and still wind up with an STS. In particular, let $\langle P', B'\rangle$ be a sub-STS of $\langle P, B\rangle$ (that is, $\langle P', B'\rangle$ is an STS and $P' \subset P$, $B' \subset B$), and let $\langle P', B''\rangle$ be an STS; then $\langle P, (B-B')\cup B''\rangle$ is an STS. We will need a slightly more general construction, called a cable by Mendelsohn. For $i \in I$ let $\langle P_i, B_i\rangle$ be an STS; moreover assume that for $i,j \in I$, if distinct $a,b \in P_i \cap P_j$, then the unique block containing a and b is the same in B_i and B_j. Let $P = \cup P_i$ and $B = \cup B_i$; then $\langle P, B\rangle$ is the <u>cable</u> formed by $\{\langle P_i, B_i\rangle \mid i \in I\}$; in general it is only a partial STS.

Let $\langle P, B\rangle$ be an STS with sub-STSs $\{\langle P_i, B_i\rangle \mid i \in I\}$; obviously, $\{\langle P_i, B_i\rangle \mid i \in I\}$ forms a cable which we denote by $\langle P', B'\rangle$. Suppose $\{\langle P_i, B_i''\rangle \mid i \in I\}$ are STSs forming a cable $\langle P', B''\rangle$. Then $\langle P, (B-B')\cup B''\rangle$ is an STS.

Let $\underline{G}$ be a finite group. A standard result from graph theory is that $\underline{G}$ can be represented as the automorphism group of a finite directed graph with no loops, isolated points, 2-cycles or triangles (3 vertices with 3 edges betweed them); denote it by <Y, S>. Define X to be $Y \dot{\cup} (S \times \{0,1\})$ and define $R \subset X^4$ by $R = \{(x, ((x,y),0), ((x,y),1), y) \mid (x,y) \in S\}$. Clearly, $\underline{G}$ is isomorphic to the automorphism group of <X, R>.

Since X is finite, we may assume that X is a free generating set of $\underline{2}^n$ where $n = |X|$. We will now use the cabling construction on $\langle P_n, B_n \rangle$ (the STS associated with $\underline{G}$) to obtain an STS whose automorphism group is isomorphic to $\underline{G}$. For this we need to use a 15-element STS, <Q, C>, which has no non-trivial automorphisms and no subsystems of order 7 (of course, each point is a subsystem of order 1 and each block is a subsystem of order 3). Since a subloop of a loop is at most half the size of the loop, these are the only subsystems of <Q, C>. Thus, every triangle (3 points not in a block) generates <Q, C>.

For $r = (r_0,r_1,r_2,r_3) \in R$, let $\langle T_r, D_r \rangle$ be the subsystem of $\langle P_n, B_n \rangle$ generated by $\{r_0,r_1,r_2,r_3\}$; it has 15 elements since the corresponding sloop is isomorphic to $\underline{2}^4$. Let <T, D> be the cable formed by $\{\langle T_r, D_r \rangle \mid r \in R\}$. For distinct r, s, t, $|T_r \cap T_s| \le 1$. Also, $\langle T_r, D_r \rangle$ is isomorphic to $\langle T_s, D_s \rangle$ via the isomorphism $\phi_{r,s}$ defined by mapping r_i to s_i.

For some fixed $r \in R$ define $\langle T_r, D_r' \rangle$ to be isomorphic to <Q, C> with $\phi_r : Q \to T_r$ being an isomorphism. Then for each $s \in R$ we define $\langle T_s, D_s' \rangle$ to be isomorphic to <Q, C> via the isomorphism $\phi_{r,s}\phi_r$. The reader should now check that $\{\langle T_s, D_s' \rangle \mid s \in R\}$ forms a cable <T, D'>. With $B_n' = (B_n - D) \cup D'$, $\langle P_n, B_n' \rangle$ is an STS.

<u>Theorem 2.1 (E. Mendelsohn [6])</u>: The automorphism group of $\langle P_n, B_n' \rangle$ is isomorphic to $\underline{G}$.

The proof is left to the reader; it is not difficult but does require care. The reader should see how all the pieces fit together, in particular why the more complicated relation $\langle X, R\rangle$ is used instead of the graph $\langle Y, S\rangle$.

Since $\underline{G}$ was an arbitrary finite group, we now know that every finite group is isomorphic to the automorphism group of a finite STS. What about other classes of block designs (such as S(2,k) where all blocks have size k)? The situation will certainly turn out to be the same as for $k = 3$, but no proofs exist at this time. The cabeling property holds for S(2,k) as well as for S(2,3). But there are two roadblocks to extending Mendelsohn's technique. The first is the existence of an analog for $\langle P_n, B_n\rangle$ which has a large automorphism group and many isomorphic small subsystems. For $k = p^m+1$ (or $k = p^m$) with p prime, the corresponding projective (affine) geometries can play the role of the $\langle P_n, B_n\rangle$. For other values of k, no such convenient designs seem available. The second aspect is the existence of an analog for $\langle Q, C\rangle$. It must be of the same size as the numerous small subdesigns of the analog of $\langle P_n, B_n\rangle$; it must be rigid (have no non-trivial automorphisms), and it must be planar (generated by every triangle). Such designs are likely extremely abundant; unfortunately, the standard constructions tend to produce designs with automorphisms or subdesigns.

3. Coding Theory as Universal Algebra.

Algebraic coding theory is a well established discipline; it should not be too surprising then that a universal algebraic approach to coding theory is possible. Significantly, this approach is not just a sterile generalization but has led to the discovery of some new perfect codes.

Given a finite set A (called an alphabet), an (n,k)-code, (α,β), over A is a pair of functions $\alpha:A^k \to A^n$ and $\beta:A^n \to A^k$ such that $\beta(\alpha(x)) = x$; α is the encoding function and β the decoding function. The code is systematic if there are n-k functions $f_1, \ldots, f_{n-k}$ such that for $\underline{a} \in A^k$, we have $\alpha(\underline{a}) = (\underline{a}, f_1(\underline{a}), \ldots, f_{n-k}(\underline{a}))$. We call the function $F = (f_1,\ldots,f_{n-k}):A^k \to A^{n-k}$ the redundancy function of (α,β). We will only be considering systematic codes. For $\underline{a},\underline{b} \in A^n$, $d(\underline{a},\underline{b})$ is the number of components in which $\underline{a}$ and $\underline{b}$ differ; (α,β) has minimum distance d if d is the largest integer such that for all $\underline{a},\underline{b} \in A^k$, $d(\alpha(\underline{a}),\alpha(\underline{b})) \geq d$.

If A is a one dimensional vector space over GF(q) and $\alpha(A^k)$ is a subspace of A^n, then (α,β) is called linear. In this case, F is a linear transformation from A^k to A^{n-k}; i.e., an $(n-k)\times k$ matrix with entries in GF(q).

This is all standard terminology and has nothing to do with universal algebra. Universal algebra enters the picture when we think of codes in terms of free algebras. Let $\underline{F}(k)$ be a free algebra with free generating set $\{x_1,\ldots,x_k\}$. An n-arc of order d in $\underline{F}(k)$ is an n-element subset, $\{x_1,\ldots,x_k,t_1,\ldots,t_{n-k}\}$, of $\underline{F}(k)$ such that every (n-d+1)-element subset of it generates $\underline{F}(k)$ (not necessarily freely). If $d = n-k+1$, then $t_1,\ldots,t_{n-k}$ are mutually orthogonal k-skeins (in particular for $k = 2$, they are mutually orthogonal latin squares). The case $d = 1$ is uninteresting.

Just as for mutually orthogonal latin squares, the existence of an n-arc in $\underline{F}(k)$ is equivalent to a set of identities being true in $\underline{F}(k)$. For instance, if we have an n-arc of order 2, $\{x_1,\ldots,x_k,t_1,\ldots,t_{n-k}\}$, then there is some term t such that $t(x_2,\ldots,x_k,t_1,\ldots,t_{n-k}) = x_1$; this is an identity in the variables $x_1,\ldots,x_k$. Given $\underline{F}(k)$, let V_k be the largest variety whose k-generated free algebra is $\underline{F}(k)$; V_k is the class of all algebras (of the same type as $\underline{F}(k)$) whose k-generated subalgebras are quotients of $\underline{F}(k)$.

Theorem 3.1 (B. Ganter): Let $\underline{B}$ be a finite algebra in V_k and let $\{x_1,\ldots,x_k,t_1,\ldots,t_{n-k}\}$ be an n-arc of order d in $\underline{F}(k)$. Define $\alpha: B^k \to B^n$ by $\alpha(\underline{a}) = (\underline{a}, t_1(\underline{a}), \ldots, t_{n-k}(\underline{a}))$ and β so that $\beta(\alpha(\underline{a})) = \underline{a}$. Then (α,β) is a systematic (n,k)-code with minimum distance at least d.

Proof: We must show that for distinct $\underline{a}, \underline{b} \in B^k$, $d \le d(\alpha(\underline{a}),\alpha(\underline{b}))$. But this follows from the fact that our n-arc has order d: every (n-d+1)-element subset, $\{g_1,\ldots,g_{n-d+1}\}$, of our n-arc generates the k projection functions. Hence, if $g_i(\underline{a}) = g_i(\underline{b})$ for $1 \le i \le n-d+1$, then $a_i = b_i$ for $1 \le i \le k$, i.e. $\underline{a} = \underline{b}$. Thus, $\alpha(\underline{a})$ and $\alpha(\underline{b})$ must differ in at least d components: $d \le d(\alpha(\underline{a}),\alpha(\underline{b}))$.

Conversely, given a systematic (n,k)-code (α,β) of minimum distance d on alphabet A, we would like to find an algebra $\underline{A} = \langle A; O\rangle$ and an n-arc of order d in $\underline{F}_A(k)$, the k-generated free algebra over $\underline{A}$, such that (α,β) is induced by $\underline{F}_A(k)$. This turns out to be quite easy. We start by thinking of $\underline{F}_A(k)$ as a set of k-ary functions on A. There is an obvious candidate for our n-arc: the k projection functions $x_1,\ldots,x_k$ and the n-k redundancy functions $f_1,\ldots,f_{n-k}$ associated with (α,β). Thus, we put the functions $f_1,\ldots,f_{n-k}$ into our set O of operations. Next, we must check to see whether our n-arc has order d (trivially, it has order ≥ 1). Let $G = \{g_1,\ldots,g_{n-d+1}\}$ be an (n-d+1)-element subset of our n-arc and suppose that $x_i \notin G$. We must ensure that G generates x_i. For this, we simply define a function $f_{G,i}: A^{n-d+1} \to A$ such that $f_{G,i}(g_1,\ldots,g_{n-d+1}) = x_i$, and add it to O. Before feeling too satisfied with ourselves, we should check to see that such a function exists. This follows from the fact that (α,β) has minimum distance d: for $\underline{a},\underline{b} \in A^k$, if $g_i(\underline{a}) = g_i(\underline{b})$ for $1 \le i \le n-d+1$, then $\underline{a} = \underline{b}$. Hence, there is a partial function $f_{G,i}$ such that for all $\underline{a} \in A^k$, $f_{G,i}(g_1(\underline{a}),\ldots,g_{n-d+1}(\underline{a})) = a_i$. Extend $f_{G,i}$ arbitrarily to a full function and we are done. Thus, O consists of the n-k redundancy functions and all the functions $f_{G,i}$.

Theorem 3.2: Every systematic (n,k)-code arises as described in Theorem 3.1.

Instead of making O so large, we could have achieved the same end by making $\underline{A}$ primal; we could merely have made O consist of one Sheffer function. Since every function on A is generated by a Sheffer function on A, all the functions we need would be available. But there is a reason why we might not want to do this. Namely, given one systematic code, we would like to generate others in some uniform manner. Given one systematic code, we know that we can form a free algebra $\underline{F}(k)$ and a variety V_k. But we also know that every finite algebra in V_k induces a systematic code with the same parameters n,k,d. The larger we make $\underline{F}(k)$, the sparser the finite models in V_k become, and so there are fewer systematic codes which we can construct from the variety.

This is a very graceful univeral algebraic result which arises by constructing certain functions. By constructing certain other functions (mere exactly, by modifying some existing functions), we can also generate some new 1-error correcting perfect codes. A code is <u>1-error correcting</u> if $d \geq 3$ (i.e. for $\underline{a}' \in A^n$ and $\underline{a} \in A^k$, if $d(\underline{a}',\alpha(\underline{a})) \leq 1$, then $\underline{a}$ is the unique element in A^k with this property). A 1-error correcting code is <u>perfect</u> if for each $\underline{a}' \in A^n$ there is a unique $\underline{a} \in A^k$ such that $d(a',\alpha(\underline{a})) \leq 1$; in this case, $\beta(\underline{a}') = \underline{a}$.

Obviously, the conditions for a 1-error correcting perfect code are quite restrictive. However, if we start with one, perhaps we can modify its redundancy functions to produce a new code. This is what Bauer, Ganter and Hergert [1] have done. We start by taking $A = \{0,1\}$. For our algebra $\underline{A}$ we take the 2-element field GF(2). Since GF(2) is primal, every function on $\{0,1\}$ can be expressed as a classical polynomial over GF(2). Since $2x = 0$ and $x^2 = x$, each polynomial is a sum of products of distinct variables.

If we start with a linear code, then our redundancy function $F = (f_1,\dots,f_{n-k})^T$ consists of n-k linear functions in $x_1,\dots,x_k$; hence, it is determined by an $(n-k)\times k$ matrix over GF(2) (its coefficient matrix) which we denote by J(F). The key to modifying F is to think of J(F) as the Jacobian of the transformation $F:\{0,1\}^k \to \{0,1\}^{n-k}$. Since every function on $\{0,1\}$ can be written as a polynomial over GF(2), we can formally define the Jacobian of every transformation $F:\{0,1\}^k \to \{0,1\}^{n-k}$, linear or not. The entries of J(F) are actually finite differences; if $(f_i)_{x_j}$ is the formal partial derivative of f_i with respect to x_j, then

$(f_i)_{x_j}(a_1,\dots,a_k) = f(a_1,\dots,a_j+1,\dots,a_k) - f(a_1,\dots,a_j,\dots,a_k)$.

This means that the jth column of J(F) applied to $(a_1,\dots,a_j,\dots,a_k)$ gives the change in the redundancy function F as we move from $(a_1,\dots,a_j,\dots,a_k)$ to $(a_1,\dots,a_j+1,\dots,a_k)$. If we assume that $F(\underline{0}) = \underline{0}$ (which causes no loss of generality), then F is uniquely determined by J(F).

Let J be an $(n-k)\times k$ matrix with entries being polynomials over GF(2). When is $J = J(F)$ for F the redundancy function of a 1-error correcting perfect code? First we must insure that J is the Jacobian of some transformation F. Let H_i be the ith column of J; it is easy to check that J is J(F) for some F if and only if for all $1 \le i,j \le k$, $(H_i)_{x_i} = \underline{0}$ and $(H_i)_{x_j} = (H_j)_{x_i}$. Supposing now that our J passes this test, our question is answered by the following result. For $\underline{v} \in \{0,1\}^m$, the weight of $\underline{v}$ is the number of non-zero components of $\underline{v}$.

Theorem 3.3 (H. Bauer, B. Ganter and F. Hergert [1]): J = J(F) is the Jacobian of a 1-error correcting perfect $(2^{r-1}, 2^r-1-r)$-code over GF(2) if and only if for every $\underline{a} \in \{0,1\}^k$, the columns of $J(F)(\underline{a})$ run through all r-tuples of weight ≥ 2.

This theorem is relatively easy to apply as the following example shows.

Example (H. Bauer, B. Ganter and F. Hergert [1]): Take r = 4 and let J^T (the transpose of J) be:

	f_1	f_2	f_3	f_4
x_1	0	0	1	1
x_2	0	1	1	0
x_3	0	1	0	1
x_4	0	1	1	1
x_5	1	0	1	$x_8+x_9+x_{10}+x_{11}$
x_6	1	0	0	1
x_7	1	0	1	$1+x_8+x_9+x_{10}+x_{11}$
x_8	1	1	$x_{10}+x_{11}$	x_5+x_7
x_9	1	1	$1+x_{10}+x_{11}$	x_5+x_7
x_{10}	1	1	x_8+x_9	$1+x_5+x_7$
x_{11}	1	1	$1+x_8+x_9$	$1+x_5+x_7$

The many symmetries involved in J make it rather easy to see that J is indeed a Jacobian matrix satisfying the conditions of Theorem 3.3. In fact, it turns out that the code corresponding to J is not equivalent to any previously known code.

4. Categorical Combinatorics.

At first glance, category theory may seem singularly inappropriate for studying combinatorics. Morphisms are the essence of category theory, but combinatorics tends to focus on structures and their substructures. When the typical morphisms are just embeddings, why bother with category theory? But there are naturally occurring examples where morphisms are not always 1-1. For instance, if our structures are totally unstructured, just sets, then every map is a morphism; moreover, some of the most fundamental enumerative combinatorics deals with sets and set mappings. Thus, there may well be a useful role for category theory in combinatorics.

One way of using category theory is via the concept of a Möbius category, introduced by Pierre Leroux [4] to provide a unified setting for Möbius inversion. Given a category $\underset{\sim}{C}$, form $A[[\underset{\sim}{C}]]$, the algebra of formal power series over commutative ring A (with the morphisms of $\underset{\sim}{C}$ being the "variables") where multiplication is given by

$$(\Sigma\alpha_f f)(\Sigma\beta_f f) = \Sigma\gamma_f f \text{ where } \gamma_f = \Sigma\alpha_{f'}\beta_{f''}$$

(the last sum is taken over all (f',f") such that f'f" = f). For this definition to make sense, we need to assume that {(f',f") | f'f" = f} is finite; such a category is called <u>decomposition finite</u>. $A[[\underset{\sim}{C}]]$ is an associative A-algebra with identity δ defined by $\delta(f) = 1$ if f is an identity morphism and $\delta(f) = 0$ otherwise. A decomposition finite category is a <u>Möbius category</u> if each morphism has only finitely many strict decompositions $f = f_1 \ldots f_n$. For a Möbius category $\underset{\sim}{C}$, if $\alpha \in A[[\underset{\sim}{C}]]$, then $\alpha = \Sigma\alpha_f f$ is invertible iff for every identity morphism e of $\underset{\sim}{C}$, α_e is invertible in A. Hence, $\zeta = \Sigma\zeta_f f$ where $\zeta_f = 1$ for all f in $\underset{\sim}{C}$ is invertible; let its inverse be given by $\mu = \Sigma\mu_f f$. Then the standard Möbius inversion formula holds: for $\alpha = \Sigma\alpha_f f$ and $\beta = \Sigma\beta_f f$,

$$\text{if } \beta_f = \Sigma\{\alpha_{f'} \mid f'f'' = f\}, \text{ then } \alpha_f = \Sigma\{\beta_{f'}\mu_{f''} \mid f'f'' = f\}.$$

Another approach, due to A. Meush, is intimately connected with the concept of a factorization of a category; consequently, this approach uses much more category theory. However, we will not go into any category theory in this paper. Interestingly, the universal algebraic concepts of direct product and subdirect product are the starting point for Meush's approach.

Let $\underset{\sim}{K}'$ be a class of finite algebras closed under finite products and subalgebras; let $\underset{\sim}{K}$ be the skeletal category associated with $\underset{\sim}{K}'$: the objects of $\underset{\sim}{K}$ consist of one member of each isomorphism class of $\underset{\sim}{K}'$ and the morphisms are all algebra homomorphisms. Let $Z(\underset{\sim}{K})$ be the free Z-module with basis the objects of $\underset{\sim}{K}$; we will define two multiplications on $Z(\underset{\sim}{K})$.

Let $Ob(\underset{\sim}{K}) = \{A_i \mid i \in I\}$; for $i,j \in I$, let $A_{p(i,j)}$ be that member of $Ob(\underset{\sim}{K})$ isomorphic to the direct product of A_i and A_j, $A_i \times A_j$. For the first multiplication, let $A_i \circ A_j = A_{p(i,j)}$ and extend to $Z(\underset{\sim}{K})$ linearly. For the second, let $r(i,j,k)$ be the number of subdirect subalgebras of $A_i \times A_j$ (i.e. subalgebras of $A_i \times A_j$ which project onto both A_i and A_j) which are isomorphic to A_k and define $A_i * A_j = \Sigma r(i,j,k)A_k$; extend to $Z(\underset{\sim}{K})$ linearly. Note that $r(i,j,k)$ is always finite and for fixed i and j, $r(i,j,k)$ is non-zero for only finitely many k. Thus, $*$ is well defined.

<u>Theorem 4.1 (A. Meush [7])</u>: $\langle Z(\underset{\sim}{K}); \circ\rangle$ and $\langle Z(\underset{\sim}{K}); *\rangle$ are isomorphic rings.

<u>Proof</u>: For $i \in I$, define $t(A_i) = \Sigma t(i,j)A_j$ where $t(i,j)$ is the number of subalgebras of A_i isomorphic to A_j; let t be the linear extension to a Z-module endomorphism of $Z(\underset{\sim}{K})$. This will be our isomorphism from $\langle Z(\underset{\sim}{K}); \circ\rangle$ to $\langle Z(\underset{\sim}{K}); *\rangle$. To see that multiplication is preserved, notice that

$$t(p(i,j),k) = \Sigma t(i,m)t(j,n)r(m,n,k).$$

Hence, $t(A_i \circ A_j) = t(A_i) * t(A_j)$. To complete the proof, we need to show that t is invertible. Let us write $t = id+u$ where id is the identity map and $u(A_i) = \Sigma\{t(i,j)A_j \mid i \neq j\}$. We can define a partial order on I by $i \leq j$ iff $t(j,i) > 0$ (i.e. iff A_i can be embedded into A_j). Then $u(A_i) = \Sigma\{t(i,j)A_j \mid j < i\}$. Since every principal order ideal of $\langle I, \leq\rangle$ is finite, u is locally nilpotent: for each $w \in Z(\underset{\sim}{K})$ there is an n such that $u^n(w) = 0$. This means that the inverse of t is $t^{-1} = \Sigma(-1)^i u^i$ (the local nilpotence of u guarantees that the sum is always finite).

Let $Hom(A_i,A_j)$ be the set of all homomorphisms from A_i to A_j and $Sur(A_i,A_j)$ the set of all onto homomorphisms from A_i to A_j; define $d(i,j) = |Sur(A_i,A_j)|$ and $c(i,j) = |Hom(A_i,A_j)|$. Define $d_i:Z(\underset{\sim}{K}) \to Z$ by $d_i(A_j) = d(i,j)$ and define $c_i:Z(\underset{\sim}{K}) \to Z$ by $c_i(A_j) = c(i,j)$, with both extended linearly. Finally, define $d:Z(\underset{\sim}{K}) \to Z^I$ by $d = (d_i)_{i\in I}$ and $c:Z(\underset{\sim}{K}) \to Z^I$ by $c = (c_i)_{i\in I}$.

<u>Theorem 4.2 (A.Meush [7])</u>: $c:\langle Z(\underset{\sim}{K}); \circ\rangle \to Z^I$ and $d:\langle Z(\underset{\sim}{K}); *\rangle \to Z^I$ are ring embeddings such that $dt = c$ and $ct^{-1} = d$.

The proof is fairly straightforward and is left as an exercise. That c is an embedding was first proved by L. Lovász in [5] where it is the basis for his results on cancellation of common direct factors in categories of finite structures. Using 4.1 and 4.2 we can compute a number of combinatorial identities. Let us write $t^{-1}(A_i) = \Sigma w(i,j)A_j$. Then it is easy to prove the following:

(a) $r(i,j,k) = \Sigma w(i,m)w(j,n)t(p(m,n),k)$.

(b) $t(p(i,j),k) = \Sigma t(i,m)t(j,n)r(m,n,k)$.

(c) $c(i,j) = \Sigma t(j,k)d(i,k)$.

(d) $d(i,j) = \Sigma w(j,k)c(i,k)$.

(e) $c(r,i)c(r,j) = c(r,p(i,j))$.

(f) $d(r,i)d(r,j) = \Sigma r(i,j,k)d(r,k)$.

Notation: To indicate a specific category $\underset{\sim}{K}$, the numerical functions r, t, c, d, w will be subscripted with the letter K.

Example 1: Finite Sets. For $1 \le i < \omega$, let A_i be an i-element set. Our category is $\underset{\sim}{S}$, the category with object set $\{A_i \mid 1 \le i < \omega\}$ and with all mappings as morphisms. Thus, $p_S(i,j) = ij$; $c_S(i,j) = j^i$; $d_S(i,j) = j!S(i,j)$ where $S(i,j)$ is the Stirling number of the second kind; $t_S(i,j) = \binom{i}{j}$; $w_S(i,j) = (-1)^{i-j}\binom{i}{j}$. The numbers $r_S(i,j,k)$ do not seem to have a common expression; they can be interpreted as the number of $i \times j$ (0,1)-matrices with at least one 1 in each row and column and exactly k 1s occurring. The identities (a)-(f) become:

$$(S_a) \qquad r_S(i,j,k) = \Sigma(-1)^{i+j-m-n}\binom{i}{m}\binom{j}{n}\binom{mn}{k}.$$

$$(S_b) \qquad \binom{ij}{k} = \Sigma\binom{i}{m}\binom{j}{n}r_S(m,n,k).$$

$$(S_c) \qquad j^i = \Sigma\binom{j}{k}k!S(i,k).$$

$$(S_d) \qquad j!S(i,j) = \Sigma(-1)^{j-k}\binom{j}{k}k^i.$$

$$(S_e) \qquad i^r j^r = (ij)^r.$$

$$(S_f) \qquad i!j!S(r,i)S(r,j) = \Sigma r_S(i,j,k)k!S(r,k)$$

$$= \Sigma(-1)^{i+j-m-n}\binom{i}{m}\binom{j}{n}\binom{mn}{k}k!S(r,k).$$

Notice that (S_e) is trivial, (S_b) obvious by counting (0,1)-matrices, (S_c) and (S_d) well known; (S_a) is the inverse of (S_b); (S_f) can be obtained by interpreting $c_S(i,j)$ and $r_S(i,j,k)$ in the combinatorial manner already mentioned.

Example 2: Finite Boolean Algebras. For $1 \le i < \omega$, let B_i be a boolean algebra with i atoms. Our category, $\underset{\sim}{B}$, has $\{B_i \mid 1 \le i < \omega\}$ as its object set with morphisms being the boolean homomorphisms. As is well known, $\underset{\sim}{B}$ is dual to the category $\underset{\sim}{S}$. Thus, $p_B(i,j) = i+j$; by duality, $c_B(i,j) = i^j$; $d_B(i,j) = j!\binom{i}{j}$; $t_B(i,j) = S(i,j)$; $w(i,j) = s(i,j)$, the Stirling number of the first kind; $r_B(i,j,k) = \binom{i}{i+j-k}\binom{j}{i+j-k}(i+j-k)!$ (by duality; combinatorially this is the number of maps from an (i+j)-set onto a k-set whose restrictions to the first i elements and to the last j elements of the (i+j)-set are 1-1). The identities (a)-(f) become:

(B_a) $\binom{i}{i+j-k}\binom{j}{i+j-k}(i+j-k)! = \Sigma s(i,m)s(j,n)S(m+n,k)$.

(B_b) $S(i+j,k) = \Sigma S(i,m)S(j,n)\binom{m}{m+n-k}\binom{n}{m+n-k}(m+n-k)!$.

(B_c) $i^j = \Sigma S(j,k)k!\binom{i}{k}$.

(B_d) $j!\binom{i}{j} = \Sigma s(j,k)i^k$.

(B_e) $r^i r^j = r^{i+j}$.

(B_f) $i!j!\binom{r}{i}\binom{r}{j} = \Sigma\binom{i}{i+j-k}\binom{j}{i+j-k}(i+j-k)!k!\binom{r}{k}$;

or $\binom{r}{i}\binom{r}{j} = \Sigma\binom{k}{i}\binom{i}{k-j}\binom{r}{k}$.

Notice that (B_e) is trivial, (B_c) and (B_d) are standard, (B_f) is well known; (B_a) and (B_b) seem rather obscure. However, if we take j = 1 in (B_b), then we get the basic recurrence relation for Stirling numbers of the second kind:

$$S(i+1,k) = S(i,k-1) + kS(i,k).$$

There is an important special case, namely when k = i. In this case, r(i,j,i) = d(i,j). This follows from the fact that each onto homomorphism $\phi: A_i \to A_j$ can be represented as $\{(a,\phi(a)) \mid a \in A_i\}$ which is a subdirect product of A_i and A_j isomorphic to A_i, and conversely. Combining equations (a) and (d) we get:

(g) $$\sum_n w(j,n)c(i,n) = \sum_{m,n} w(j,n)w(i,m)t(p(m,n),i).$$

Because t and t^{-1} are inverses of each other, for any sequences $\{a_i\}_{i\in I}$ and $\{b_i\}_{i\in I}$ of real numbers, we have: $a_i = \Sigma t(i,j)b_j$ for all i if and only if $b_i = \Sigma w(i,j)a_j$ for all i. Applying this inversion to (g) we get:

(h) $$c(i,n) = \Sigma w(i,m)t(p(m,n),i).$$

For our categories $\underset{\sim}{S}$ and $\underset{\sim}{B}$, we get:

(S_h) $$j^i = \Sigma(-1)^{i-m}\binom{i}{m}\binom{mj}{i}.$$

(B_h) $$i^j = \Sigma s(i,m)S(m+j,i).$$

Equation (B_h) is particularly appealing because of its simplicity. Yet it is at best obscure and perhaps previously unknown.

Let us return to the rings $\langle Z(\underset{\sim}{K}); \circ\rangle$ and $\langle Z(\underset{\sim}{K}); *\rangle$. Clearly, $\langle Z(\underset{\sim}{K}); \circ\rangle$ is generated by $D(\underset{\sim}{K})$, those $A_j \in \underset{\sim}{K}$ which are directly irreducible. In fact, if $\underset{\sim}{K}$ has unique direct factorization, then $\langle Z(\underset{\sim}{K}); \circ\rangle$ is freely generated by $D(\underset{\sim}{K})$. What about $\langle Z(\underset{\sim}{K}); *\rangle$? Since it is isomorphic to $\langle Z(\underset{\sim}{K}); \circ\rangle$, the same is true, but with $t(D(\underset{\sim}{K}))$, the image of $D(\underset{\sim}{K})$ under t, as (free) generating set. It is natural to ask whether $D(\underset{\sim}{K})$ itself is a (free) generating set for $\langle Z(\underset{\sim}{K}); *\rangle$.

<u>Lemma 4.3</u>: $\langle Z(\underset{\sim}{K});*\rangle$ is generated by $D(\underset{\sim}{K})$.

<u>Proof</u>: Induct on $|A_n|$; it suffices to show that each reducible A_n is a polynomial in some finite subset D of $D(\underset{\sim}{K})$ with $|A_i| < |A_n|$ for each $A_i \in D$. For this, use the equation

$$A_{p(i,j)} = A_i * A_j - \Sigma\{r(i,j,k)A_k \mid k < p(i,j)\}$$

and note that if $k < p(i,j)$, then $|A_k| < |A_{p(i,j)}|$.

<u>Theorem 4.4 (A. Meush [7])</u>: If $\underset{\sim}{K}$ has unique direct factorization, then $D(\underset{\sim}{K})$ is a free generating set for $\langle Z(\underset{\sim}{K}); *\rangle$.

<u>Proof</u>: For $i,j \in I$, define A_i <u>precedes</u> A_j if (A_i,A_j) is in the transitive closure of the union of the relations (i) that A_s is properly embeddable in A_t and (ii) that A_s is a proper direct factor of A_t. Note that if A_i precedes A_j, then $|A_i| < |A_j|$;

thus, this relation is a partial order on I. The assumption that $\underset{\sim}{K}$ has unique direct factorization implies that every finitely generated order ideal of I under this partial order is finite (if our type is finite, then we do not need unique factorization at this point). Let C be a finite subset of $D(\underset{\sim}{K})$ and D the members of $D(\underset{\sim}{K})$ belonging to the order ideal of I generated by C. It is easily seen that the subring of $\langle Z(\underset{\sim}{K}); *\rangle$ generated by D is the same as the subring of $\langle Z(\underset{\sim}{K}); *\rangle$ generated by $t(D) = \{t(A_i) \mid A_i \in D\}$; denote this ring by $\langle Z(D); *\rangle$. But as $t(D(\underset{\sim}{K}))$ generates $\langle Z(\underset{\sim}{K}); *\rangle$ freely, $t(D)$ generates $\langle Z(D); *\rangle$ freely. Hence there is an onto endomorphism ϕ_D of $\langle Z(D); *\rangle$ defined by $\phi_D(t(A_i)) = A_i$ for all $A_i \in D$. But $\langle Z(D); *\rangle$, being finitely generated, is Noetherian, and in a Noetherian ring every onto endomorphism is an isomorphism. Hence, if we define ϕ on $\langle Z(\underset{\sim}{K}); *\rangle$ by $\phi(t(A_i)) = A_i$ for all $A_i \in D(\underset{\sim}{K})$, then ϕ extends to an isomorphism of $\langle Z(\underset{\sim}{K}); *\rangle$, proving the theorem.

<u>Corollary 4.5</u>: If $\underset{\sim}{K}$ has unique direct factorization, then every A_i is expressible as a unique polynomial (with integer coefficients and zero constant term) in the members of $D(\underset{\sim}{K})$ which precede or equal A_i.

Recall our embedding $d:\langle Z(\underset{\sim}{K}), *\rangle \to Z^I$. For each $n \in I$, $d(A_n)$ is the function $d(\_,n):I \to Z$ where $d(\_,n)(i) = d(i,n)$. Thus, we see that for each $n \in I$, there is a unique polynomial, P_n, (with integer coefficients and zero constant term) in the directly irreducible A_i which precede or equal A_n (say, $A_1,\ldots,A_k$) such that $d(\_,n) = P_n(d(\_,1),\ldots,d(\_,k))$.

For $\underset{\sim}{S}$, A_n is directly irreducible iff n is prime and A_i precedes A_j iff $i < j$. Since $d_S(i,j) = j!S(i,j)$, we see that for each n there is a unique polynomial P_n (with rational coefficients and zero constant term) in the primes $\leq n$ (say $p_1,\ldots,p_k$) such that $S(\_,n) = P_n(S(\_,p_1),\ldots,S(\_,p_k))$. For example, $S(\_,4) = \frac{1}{6}S(\_,2)^2 - S(\_,3) - \frac{1}{6}S(\_,2)$. The existence of such a polynomial can be inferred from equation (S_d). For $\underset{\sim}{B}$, B_1 is the only directly irreducible. The reader can check that the corresponding polynomial is well known.

References

[1] H. Bauer, B. Ganter and F. Hergert, Algebraic techniques for nonlinear codes, Fachbereich Mathematik, TH Darmstadt, Preprint-Nr. 609.

[2] T. Evans, Universal algebra and Euler's officer problem, Amer. Math. Monthly, 86(1979), 466-473.

[3] T. Evans, Universal-algebraic aspects of combinatorics, Universal algebra (Esztergom, 1977), Colloq. Math. Soc. János Bolyai, 29(1982), 241-266.

[4] P. Leroux, Les catégories de Möbius, Cahier de top. et géom. diff. 16(1975), 280-282.

[5] L. Lovász, Operations with structures, Acta Math. Sci. Hungar. 18(1967), 321-328.

[6] E. Mendelsohn, On the groups of automorphisms of Steiner triple and quadruple systems, J. Comb. Theory (A) 25(1978), 97-104.

[7] A. Meush, Categorical combinatorics, Ph.D. Thesis, University of Manitoba, Winnipeg, 1980.

Universal Algebra and its Links with Logic,
Algebra, Combinatorics, and Computer Science
Proc. "25. Arbeitstagung über Allgemeine Algebra", Darmstadt 1983
P. Burmeister et al. (eds.)

168 – 172

UNIVERSAL ALGEBRA AND COMBINATORICS – A SERIES OF PROBLEMS

A. Beutelspacher

This short note is not intended to be a survey on already known applications of universal algebra to combinatorics. In accordance with the idea of the whole conference also this contribution deals with the future. Being a non-expert in universal algebra I shall simply state some of the most important problems in finite geometry and/or combinatorics for which one could possibly hope that (universal) algebra can be of some help.

1. LATIN SQUARES

1.1 For a given positive integer s, determine the maximal number $N(s)$ of mutually orthogonal latin squares of order s.

The exact value of $N(s)$ is known only when s is a prime-power or $s = 1,2,6$.

Since a latin square is (essentially) nothing else but the multiplication table of a quasigroup, it is not surprising that questions about latin squares can be translated into questions about quasigroups and vice versa. Indeed, a partial (negative) answer to the EULER conjecture ("$N(s) = 1$, if $s \equiv 2 \pmod 4$") has been obtained by means of universal algebra (see EVANS [4]).

It is easy to verify that $N(s) = s-1$ is equivalent to the existence of a projective plane of order s. So, a particular case of 1.1 is

1.2 For which positive integers s does there exist a projective plane of order s?

Any known projective plane has prime-power order. On the other hand, the theorem of BRUCK and RYSER [2] excludes an infinite family of values, for example all integers $s \equiv 6 \pmod 8$. The first number in doubt is $s = 10$. The following special case of 1.2 seems already to be an extremely hard problem:

1.3 Does there exist a projective plane of order 10?

2. DESIGNS

A t-(v,k,λ) *design* is an incidence structure D of *points*, *blocks* and *incidences* with the following properties:

(1) D has exactly v points.

(2) Any block of D is incident with precisely k points.

(3) Any set of t points is incident with exactly λ blocks.

If $\lambda = 1$, these structures are also called *Steiner Systems*. An introduction to (the elementary part of) design theory can be found in [1]. The main question in this field is of course

2.1 Given four positive integers t,v,k and λ. Does there exist a t-(v,k,λ) design?

Using some obvious equations among the parameters of a design, one gets a number of divisibilty conditions, the so-called *necessary conditions*. Wilson [12] has shown that for $t = 2$ the necessary conditions are "asymptotically sufficient". This means: For large values of v, the necessary conditions are sufficient. Since a projective plane of order s is the same as a 2-$(s^2+s+1,s+1,1)$ design, question 1.2 is included in 2.1.

Well known designs are the finite projective and affine spaces and the WITT designs, on which the MATHIEU groups act as automorphism groups. Two of these designs are 5-designs. Recently a non-trivial 6-design has been found, while there is no non-trivial t-(v,k,λ) design known for $t>6$.

2.2 Are there non-trivial t-designs with $t>6$?

The connection between block designs (or even more general combinatorial structures) and universal algebra have been studied thoroughly. We mention a well-known example. If D is a *Steiner triple system* (i.e. a 2-$(v,3,1)$ design), then we can define a bina-nary operation o on the point set of D by

$$x \circ y = \begin{cases} x, & \text{if } x = y \\ \text{the third point on the block through } x \text{ and } y, \\ \text{if } x \neq y. \end{cases}$$

This defines a groupoid with the property that the 2-generated subalgebras are the blocks of D.
But this observation serves only as the starting point for very elaborated investigations. We refer to [6] and the literature quoted there.

3. EMBEDDINGS

The general idea in this part of combinatorics is *any partial X is embeddable in a finite X*. For example: Any partial Steiner quadruple system can be embedded in a finite Steiner quadruple system [5].

3.1 Is there a similar theorem for a larger class of designs?

3.2 Structures which "look like" a projective plane minus " a few" points and lines are in fact such structures. For example, the pseudo-complements of a point, a line, a point-line pair [3], two lines [10], or a pencil of lines [7] have been determined. Here, many problems are open.

4. RAMSEY NUMBERS

We consider only the easiest case. Let r and s be positive integers. By $R(s,t)$ we denote the smallest integer v such that any graph G with at least v vertices satisfies at least one of the following conditions:

(i) G contains a complete graph K_s as a subgraph.
(ii) The complement $\overline{G}$ of G has a subgraph isomorphic to K_t.

The theorem of RAMSEY [8] says that these numbers $R(s,t)$ exist, but only very few of them have been determined exactly.

4.1 Give better upper bounds for $R(s,t)$.

4.2 In order to improve the lower bounds, construct new examples of graphs.

5. CODES

By the theorem of SHANNON [9] we know that ther exist "arbitrarily good" codes.

5.1 Construct these good codes.

5.2 Construct codes which are good in SHANNON's sense and which have also good decoding properties.

6. FOUR COLOURS

6.1 Is there a "short" proof of the four colour theorem?

In particular: Can you do it by solving the following (harder) problem (see TUTTE [11]):

Let $P = PG(d,2)$ be the projective space of dimension d and order two. A *tangential 2-block* of P is a set $\mathscr{b}$ of points in P with the following properties:

(0) $\mathscr{b}$ spans P.

(1) Any subspace of codimension 2 of P intersects $\mathscr{b}$ in at least one point.

(2) Through any subset c of $\mathscr{b}$ spanning a subspace whose dimension is at most $d-2$, ther is a *tangent*, i.e. a $(d-2)$-dimensionsional subspace of P intersecting $\mathscr{b}$ precisely in the points of c.

6.2 Show that there exist exactly three tangential 2-blocks: the FANO-, the DESARGUES- and the PETERSEN-block.

Of course, tangential blocks are very interesting geometric structures in their own right. Therefore, we have an even more difficult problem:

6.3 Determine all tangential k-blocks in $PG(d,q)$.

REFERENCES

[1] A. Beutelspacher: Einführung in die endliche Geometrie I. Bibliographisches Institut, Mannheim 1982.

[2] R.H. Bruck and H.J. Ryser: The nonexistence of certain finite projective planes. Cand. J. Math. 1 (1949), 88-93.

[3] P. Dembowski: Semiaffine Ebenen. Arch. Math. 13 (1962), 120-131.

[4] T. Evans: Universal algebra and Euler's officer problem. Amer. Math. Monthly 86 (1979), 466-473.

[5] B. Ganter: Finite partial quadruple systems can be finitely embedded. Discrete Math. 10 (1974), 397-400.

[6] B. Ganter and H. Werner: Co-Ordinatizing Steiner Systems. Ann. of Discr. Math. 7 (1980) 3-24.

[7] R.C. Mullin and S.A. Vanstone: A generalization of a theorem of Totten. J. Austral. Math. Soc. A 22 (1976), 494-500.

[8] F.P. Ramsey: On a problem of formal logic. Proc. Lond. Math. Soc. (2) 30 (1930), 264-286.

[9] C.E. Shannon: A mathematical theory of communications. Bell Syst. Techn. J. 27 (1948), 379-423, 623-656.

[10] J. Totten: Embedding the complement of two lines in a finite projective plane. J. Austral. Math. Soc. A 22 (1976), 27-34.

[11] W.T. Tutte: On the algebraic theory of graph colorings. J. Combinat. Theory 1 (1966), 15-50.

[12] R. Wilson: An existence theory for pairwise balanced designs I, II,III. J. Combinat. Theory (A) 13, 220-245, 13, 246-273, 18, 71-79.

Albrecht Beutelspacher
Fachbereich Mathematik
der Universität Mainz
Saarstraße 21
D-6500 Mainz
Federal Republic of Germany

Universal Algebra and its Links with Logic,
Algebra, Combinatorics, and Computer Science
Proc. "25. Arbeitstagung über Allgemeine Algebra", Darmstadt 1983
P. Burmeister et al. (eds.)

173 – 182

ON GROUP UNIVERSALITY AND HOMOGENEITY

M. Funk, O. H. Kegel, K. Strambach

In our forthcoming paper we deal with the question which assumptions have to be satisfied by a class $\mathfrak{C}$ of algebraic or geometric structures so that it is possible to prescribe the automorphism group $\mathrm{Aut}(S)$ for some structure $S \in \mathfrak{C}$, or to find a structure $T \in \mathfrak{C}$ with a transitive action of $\mathrm{Aut}(T)$ on a given family of substructures:

Theorem 1 (group universality): Let $\mathrm{Th}(\Sigma)$ be an inductive theory with positive $\forall\exists$-axioms. Assume that $\mathrm{Th}(\Sigma)$ has the amalgamation property for partial models and admits Σ-hyperfree extensions. Assume further that there exists a family $\{\mathbf{J}_\nu\}_{\nu \in \mathbb{N}}$ of pairwise non-embeddable Σ-confined partial models $\mathbf{J}_\nu$ of $\mathrm{Th}(\Sigma)$. Then for any given abstract group G and any given (partial) model $\mathbf{L}$ of $\mathrm{Th}(\Sigma)$ there exists a model $\mathbf{J}$ containing a (partial) submodel isomorphic to $\mathbf{L}$ with $\mathrm{Aut}(\mathbf{J}) \cong G$.

Theorem 2 (homogeneity): Let $\mathrm{Th}(\Sigma)$ be an inductive theory with positive $\forall\exists$-axioms. Assume that $\mathrm{Th}(\Sigma)$ has the amalgamation property for partial models and admits Σ-hyperfree extensions. Then for every family $\{\mathbf{K}_\iota\}_{\iota \in I}$ of pairwise non-isomorphic models $\mathbf{K}_\iota$ of $\mathrm{Th}(\Sigma)$ there exists a model $\mathbf{K}$ of $\mathrm{Th}(\Sigma)$ containing submodels isomorphic to $\mathbf{K}_\iota$ for every $\iota \in I$ and such that for any two embeddings ξ, ξ' of $\mathbf{K}_\iota$ into $\mathbf{K}$ there is an automorphism η of $\mathbf{K}$ with $\xi\eta = \xi'$.

Forgetting about the hypotheses for a moment and only looking at the conclusions of these two theorems we'll remind us of a lot of papers presenting similar results, which are stated and proved case by case in different special languages, but partially based on the same ideas (cf. section 3 of the bibliograpgy).

The aim of our paper is to show that most of these results are in fact corollaries to a few, rather general theorems like the two mentioned above. In this way, we also obtain new results of this kind for theories which, up to now, had not been studied from this point of view: We only have to convince ourselves that the hypotheses are satisfied by the theory in question.

This essay is mainly concerned with the methods we use, namely <u>amalgamation</u> and <u>hyperfree extensions</u> of partial models. Finally, we shall give a rough description of our constructions to prove group universality and homogeneity.

1. Let $\Gamma = (V, E)$ be a graph, $E \subseteq \binom{V}{2}$, and let $\{\mathfrak{I}_v\}_{v \in V}$ and $\{\mathfrak{J}_e\}_{e \in E}$ be two families of partial models indexed by the vertices and the edges of Γ, respectively. Assume further that for each edge $e = \{u,v\}$ of Γ two embeddings $\varepsilon_u^e : \mathfrak{J}_e \longrightarrow \mathfrak{I}_u$ and $\varepsilon_v^e : \mathfrak{J}_e \longrightarrow \mathfrak{I}_v$ are given. With these data we shall define a new structure $\mathfrak{A}_\Gamma$ for the language $\mathcal{L}$, the <u>amalgam</u> of the <u>supports</u> $\mathfrak{I}_v$, $v \in V$, with respect to the <u>joint embeddings</u> ε_u^e, ε_v^e, $e = \{u,v\} \in E$.

For elements x, x' of supports $\mathfrak{I}_v$, $v \in V$, we shall write $x \sim x'$ if either $x = x'$ or if for $x \in I_v$, $x' \in I_{v'}$ there is a path $v = v_o, v_1, \dots, v_{n-1}, v_n = v'$ in Γ so that

$$x' \in x\,(\varepsilon_{v_o}^{\{v_o,v_1\}})^{-1} (\varepsilon_{v_1}^{\{v_o,v_1\}}) \dots (\varepsilon_{v_{n-1}}^{\{v_{n-1},v_n\}})^{-1} (\varepsilon_{v_n}^{\{v_{n-1},v_n\}})$$

Clearly, the relation $\sim$ is an equivalence relation on the sets of elements belonging to the same sort. Take the set of all $\sim$-equivalence classes as universe of $\mathbf{A}_\Gamma$. Finally, if R is an n-ary relation symbol of $\mathcal{L}$, define the relation R between elements $x_1, \ldots, x_n \in \mathbf{A}_\Gamma$, if and only if there exist a vertex $v \in V$ and representatives $x_1^v, \ldots, x_n^v \in \mathbf{I}_v$ for the equivalence classes $x_1, \ldots, x_n$ so that $R(x_1^v, \ldots, x_n^v)$ holds in $\mathbf{I}_v$ (cf. (KS 73)).

We may exchange the rôles played by vertices and edges; the new structure $\mathbf{A}_\Gamma$ for the language $\mathcal{L}$ is then called a <u>coamalgam</u>.

In general, the (co-) amalgam $\mathbf{A}_\Gamma$ of a family of partial models need not be again a partial model (cf. e.g. (KS 73), p. 383). In certain favourable situations, however, we can attack the question whether $\mathbf{A}_\Gamma \in \mathrm{Mod}(\mathrm{Th}(\Sigma)_\forall)$.

Consider the Skolemization Σ^S of Σ and the Σ-Skolemization $\mathcal{L}^\Sigma$ of $\mathcal{L}$ defined recursively by the following clauses:

(1) $\Sigma \subseteq \Sigma^S$ and $\mathcal{L} \subseteq \mathcal{L}^\Sigma$

(2) Let $\psi = (\exists x)\varphi$ be a formula of $\mathcal{L}^\Sigma$ with free variables $x_1, \ldots, x_n$ so that $\forall x_1, \ldots, x_n (\exists x) \varphi(x_1, \ldots, x_n, x)$ is a sentence of Σ^S. Then add a new n-ary function symbol F_ψ to $\mathcal{L}^\Sigma$ and define
$\forall x_1, \ldots, x_n \varphi(x_1, \ldots, x_n, F_\psi(x_1, \ldots, x_n)) \in \Sigma^S$ (cf. (Pa)).

Call a partial submodel $\mathbf{J}$ of a partial model $\mathbf{I}$ <u>Σ-closed</u> in $\mathbf{I}$, if the images of the functions F_ψ, applied to elements of $\mathbf{J}$, lie already in $\mathbf{J}$, as far as they do exist in $\mathbf{I}$. Finally, we say that $\mathrm{Th}(\Sigma)$ has the <u>amalgamation property for partial models</u> (APPM), if every (co-) amalgam of <u>two</u> partial models with <u>at least one</u> Σ-closed joint embedding is again a partial model.

In the light of the preceding definitions the reduction of the problem mentioned above is given by the following

<u>Theorem</u>: Assume that $\mathrm{Th}(\Sigma)$ has the APPM and let $\mathbf{A}_\Gamma$ be the (co-) amalgam of a family of partial models with Σ-closed joint embeddings.

Then $\mathbf{A}_\Gamma \in \text{Mod}(\text{Th}(\Sigma)_\forall)$, if

Γ is a tree or

$\mathbf{A}_\Gamma$ is <u>correct</u> (i.e. the images of any two joint embeddings do not intersect, except in representatives of constant symbols of $\mathcal{L}$, and the union of the images of all the joint embeddings into any support $\mathbf{J}$ of $\mathbf{A}_\Gamma$ is Σ-closed in $\mathbf{J}$, too).

For technical reasons we must also presuppose that $\text{Th}(\Sigma)$ is inductive and that the $\forall\exists$-axioms of Σ are positive.

On the whole, this reduction turns out to be sufficient for our purpose, since all the constructions we perform to show group universality and homogeneity are based on correct coamalgams and on amalgams over trees, respectively. On the other hand, it is quite easy to prove case by case that most theories for which we want to use our constructions have the APPM.

But there are still some well-known theories for which the APPM does not hold, as in the case of Laguerre planes: Take just two copies of the (classical) Laguerre plane $\mathbf{L}_\mathbb{C}$ over $\mathbb{C}$ and choose the canonical embedding of the real Laguerre plane $\mathbf{L}_\mathbb{R}$ into $\mathbf{L}_\mathbb{C}$ as joint embeddings, which are of course Σ-closed. Then every pair of circles not intersecting one another in $\mathbf{L}_\mathbb{R}$ has four different points of intersection in the corresponding amalgam $\mathbf{A}_\Gamma$. Thus $\mathbf{A}_\Gamma$ is no longer a partial Laguerre plane.

2. By amalgamation one at best obtains partial models. So we must provide a procedure to embed a partial model $\mathbf{J}$ into some model of $\text{Th}(\Sigma)$. Since we always perform amalgamations in order to impose certain properties on the automorphism group of the amalgam, we need a procedure that extends automorphisms of $\mathbf{J}$ uniquely; concerning group universality we even need a procedure that does not introduce any new automorphisms except those extending automorphisms of $\mathbf{J}$.

We shall manage this generalizing the notion of <u>hyperfree extensions</u> of <u>confined structures</u> (cf. (Ha 43)).

Let $\mathbb{M}$, $\mathbb{N}$ be partial models and assume $\emptyset \neq \mathbb{N} \subseteq \mathbb{M}$. Call the pair $(\mathbb{M}, \mathbb{N})$ a <u>Σ-free construction</u>, if for every embedding $\alpha : (\mathbb{M} \setminus \mathbb{N}) \longrightarrow \mathbb{S}$ into some model $\mathbb{S}$ of $Th(\Sigma)$ there exists a Σ-homomorphism $\alpha' : \mathbb{M} \longrightarrow \mathbb{S}$ such that the α'-image of $\mathbb{N}$ in $\mathbb{S}$ is uniquely determined and that the diagramm

$$\begin{array}{ccc} \mathbb{M} \setminus \mathbb{N} & \xrightarrow{\alpha} & \mathbb{S} \\ {\scriptstyle \alpha_o} \downarrow & \nearrow_{\alpha'} & \\ \mathbb{M} & & \end{array}$$

with the canonical embedding $\alpha_o : (\mathbb{M} \setminus \mathbb{N}) \longrightarrow \mathbb{M}$ is commutative. (For the notion of Σ-homomorphism see (Gr 79), p. 3o5.)

Let $\mathbb{I}$, $\mathfrak{J}$ be partial models and assume $\emptyset \neq \mathfrak{J} \subseteq \mathbb{I}$. At first, we build a generalized flag graph $\Delta(\mathbb{I})$ of $\mathbb{I}$ (cf. (Ti 77)). As the vertices of $\Delta(\mathbb{I})$ choose the elements of $\mathbb{I}$ not representing constant symbols of $\mathcal{L}$ and define the edges of $\Delta(\mathbb{I})$ by the following condition: vertices a_i , $i = 1,\ldots,n$, form a complete subgraph of $\Delta(\mathbb{I})$, by definition, if and only if there exist k representatives $c_j \in \mathbb{I}$ of constant symbols, $j = 1,\ldots,k$, for some $k \in \mathbb{N}_o$ and an m-ary relation symbol R of $\mathcal{L}$ for some $m \geq k + n$ so that $R(\sigma(a_i,c_j))$ holds in $\mathbb{I}$ for an m-tupel $\sigma(a_i,c_j)$ where each a_i and c_j appears at least once. Now we define a new partial submodel $\mathfrak{J} \subseteq \mathfrak{J}_{\mathbb{I}} \subseteq \mathbb{I}$, called the <u>internal construction</u> of $\mathfrak{J}$ in $\mathbb{I}$, with universe

$$J_{\mathbb{I}} = \{ x \in \mathbb{I} \;/\; \text{distance}(x,\mathfrak{J}) \leq 1 \;\text{ in } \Delta(\mathbb{I}) \} \;.$$

Actually, both Σ-free constructions and internal constructions had to be defined in a somewhat more sophisticated manner in order to avoid some trouble caused by transitive relations.

Finally, we call $\mathfrak{J}$ <u>Σ-hyperfree</u> in $\mathbb{I}$, if there exists a Σ-free construction $(\mathbb{M}, \mathbb{N})$ with embeddings $\mathfrak{J} \longrightarrow \mathbb{N}$ and $\mathfrak{J}_{\mathbb{I}} \longrightarrow \mathbb{M}$

so that the diagramm

$$\begin{array}{ccc} \mathfrak{J} & \longrightarrow & \mathfrak{N} \\ \downarrow & & \downarrow \\ \mathfrak{J}_{\mathfrak{I}} & \longrightarrow & \mathfrak{M} \end{array}$$

with canonical embeddings $\mathfrak{J} \longrightarrow \mathfrak{J}_{\mathfrak{I}}$ and $\mathfrak{N} \longrightarrow \mathfrak{M}$ is commutative. If $\mathfrak{I}$ is finite and does not contain any Σ-hyperfree partial submodel, $\mathfrak{I}$ is called Σ-<u>confined</u>. Infinite unions of finite Σ-confined partial models are called Σ-<u>confined</u>, too.

Now let $Th(\Sigma)$ be inductive and $\mathfrak{I}$ be a partial model of $Th(\Sigma)$. Consider direct limits of partial models $\mathfrak{I}_\gamma$ with $\mathfrak{I}_o = \mathfrak{I}$ and $\mathfrak{I}_\gamma \subseteq \mathfrak{I}_{\gamma+1}$ so that $\mathfrak{I}_{\gamma+1} \setminus \mathfrak{I}_\gamma$ can be covered by partial models which are Σ-hyperfree in $\mathfrak{I}_{\gamma+1}$. If one has

$$F(\mathfrak{I}) := \bigcup_{\gamma<\omega} \mathfrak{I}_\gamma \in Mod(Th(\Sigma)) ,$$

call F a Σ-hyperfree extension process.

With these notions we obtain the required procedure not permitting any new automorphism:

<u>Lemma</u>: Let F be a Σ-hyperfree extension process for an inductive theory $Th(\Sigma)$ and let $\mathfrak{I}$ be a Σ-confined partial model of $Th(\Sigma)$. Then every automorphism of $F(\mathfrak{I})$ induces an automorphism of $\mathfrak{I}$.

Again it turns out to be quite easy to define Σ-hyperfree extension processes case by case for most theories for which we want to use our constructions. If in $Mod(Th(\Sigma))$ there is involved a rather strong identity like associativity, we do not succeed in finding such a Σ-hyperfree extension process.

3. To prove group universality assuming the hypotheses of theorem 1 we first deal with the case $G = \{1\}$, i.e. we construct an embedding of a given (partial) model $\mathfrak{I}$ into a rigid one. To perform this we reformulate the recursive construction presented by A. Schleier-

macher and O. H. Kegel in (KS 73), p. 392-394, in the language of projective planes. To confine attention to the essential situation, assume $|\mathbf{I}| < \aleph_0$, too.

Partitioning the family $S = \{\mathbf{J}_\nu\}_{\nu \in \mathbb{N}}$ into $\aleph_0$ subfamilies S_μ, $\mu \in \mathbb{N}$, each of cardinality $\aleph_0$, and starting with $\mathbf{A}_0 := \mathbf{I}$ we reiterate amalgamation between $\mathbf{A}_\mu$ and the members of S_μ as supports with the elements of $\mathbf{A}_\mu$ as joints and Σ-hyperfree extension to obtain $\mathbf{A}_{\mu+1}$. Thus we get an ascending sequence of models $\mathbf{A}_\mu$ so that for $\nu \leq \mu$ one has:

$\mathbf{A}_\nu \subsetneq \mathbf{A}_\mu$, and every automorphism of $\mathbf{A}_\mu$ induces the identity in $\mathbf{A}_\nu$.

The rigid model we want is $\mathbf{I}_S := \bigcup_{\mu<\omega} \mathbf{A}_\mu$. In particular, $\mathbf{I}_S$ turns out to be Σ-confined.

For a group $G \neq \{1\}$ we start with a graph Γ such that $\mathrm{Aut}(\Gamma) \cong G$. (The existence of such a Γ is guaranteed by a series of papers beginning with the famous one by R. Frucht (Fr 38), see section 2 of the bibliography.) Then we use the rigid Σ-confined model $\mathbf{I}_S$ to blow up the edges of Γ: At first we choose two disjoint (partial) submodels $\mathbf{A}, \mathbf{B} \subseteq \mathbf{I}_S$ (mostly just two points) and amalgamate two copies of $\mathbf{I}_S$ with common joint $\mathbf{B}$ to obtain a Σ-confined partial model $\mathbf{K}$, which has precisely one non-trivial involution exchanging the partial submodels $\mathbf{I}_S \setminus \mathbf{B}$ and fixing $\mathbf{B}$. Then we build the coamalgam $\mathbf{A}_\Gamma$ over Γ with copies of $\mathbf{K}$ as supports and the induced embeddings of $\mathbf{A}$ into $\mathbf{K}$ as joint embeddings. $\mathbf{A}_\Gamma$ is a Σ-confined partial model with $\mathrm{Aut}(\mathbf{A}_\Gamma) \cong G$. Finally, we use Σ-hyperfree extension to get a model $F(\mathbf{A}_\Gamma)$ of $\mathrm{Th}(\Sigma)$ with $\mathrm{Aut}(F(\mathbf{A}_\Gamma)) \cong G$.

5. The construction to show homogeneity is a rather complicated recursive combination of creating precisely one orbit for some submodel isomorphic to $\mathbb{K}_\iota$, $\iota \in I$, and Σ-hyperfree extension. Since this construction is, in fact, a rewrite of the corresponding one by A. Schleiermacher and O. H. Kegel (KS 73), we shall only deal with the basic construction, which furnishes a transitive action on a certain set of partial submodels:

<u>Lemma</u>: Let $\mathbb{J}$ be a partial model and $\mathfrak{J}$ a Σ-closed partial submodel of $\mathbb{J}$. There exists a partial model $\mathbb{K}$ containing $\mathbb{J}$ as a Σ-closed partial submodel, such that the automorphism group of $\mathbb{J}$ is a subgroup of the automorphism group of $\mathbb{K}$ and such that for every pair ξ, ξ' of Σ-closed embeddings of $\mathfrak{J}$ into $\mathbb{J}$ there is an automorphism η of $\mathbb{K}$ with $\xi\eta = \xi'$.

Let X be the set of all ordered pairs of distinct Σ-closed embeddings of $\mathfrak{J}$ into $\mathbb{J}$. Partition the set X arbitrarily into two subsets Y, Y' so that $(\xi, \xi') \in Y$ implies $(\xi', \xi) \in Y'$. Denote by Φ the free group freely generated by the set Y, and interpret Y' as the subset Y^{-1} of Φ. As vertices of a graph Γ we choose the elements of Φ. By definition, the pair (f,f') of vertices of Γ represents an edge of Γ, if there is an element $x \in X$ so that $fx = f'$. It is easy to check that Γ is a tree.

To every vertex v of Γ associate a copy $\mathbb{J}_v$ of $\mathbb{J}$ together with a fixed isomorphism $\beta_v : \mathbb{J} \longrightarrow \mathbb{J}_v$. For every edge $e = (f,f')$ of Γ put $\mathfrak{J}_e = \mathfrak{J}$ and $\varepsilon_f^e = \xi' \beta_f$ and $\varepsilon_{f'}^e = \xi \beta_{f'}$ where $fx = f'$ for some $x = (\xi, \xi') \in X$. Then the amalgam $\mathbb{K}$ of the family $\{\mathbb{J}_f\}_{f \in \Phi}$ of supports with the joint embeddings ε_f^e, $\varepsilon_{f'}^e$ is again a partial model and there exist canonical embeddings of the $\mathbb{J}_f$, $f \in \Phi$, onto Σ-closed partial submodels of $\mathbb{K}$.

The group Φ acts as a regular group of automorphisms on the graph Γ by $v^f = f^{-1}v$ for every $f, v \in \Phi$. This action of Φ on Γ carries over to an action of Φ on the family of supports, if one puts $\beta_v f = \beta_{f^{-1}v}$ and $\xi_v^e f = \xi_{f^{-1}v}^{f^{-1}e}$. Identifying $\mathbb{I}_v$ with its canonical image in $\mathbb{K}$, it is apparent from the definition of the amalgam that f in fact acts as an automorphism of $\mathbb{K}$. Identifying $\mathbb{I}$ with the Σ-closed partial submodel $\mathbb{I}_1$ of $\mathbb{K}$ via β_1, one has for the pair ξ, ξ' of distinct Σ-closed embeddings of $\mathfrak{J}$ into $\mathbb{I}_1$ that $\xi x = \xi'$ where $\eta = x = (\xi, \xi')$ is an automorphism of $\mathbb{K}$ as required. (For further details see (KS 73), p. 387-390.)

6. Bibliography

(1) Model theory and universal algebra

J. Barwise (ed.), Handbook of Mathematical Logic, Amsterdam-New York - Oxford, North Holland 1977.

C. C. Chang and H. J. Keisler, Model Theory, Amsterdam - London, North Holland 1973.

Gr 79 G. Grätzer, Universal Algebra, 2nd edition, New York - Heidelberg - Berlin, Springer 1979.

Pa A. Pasini, On the free Σ-structures, preprint.

(2) Auxiliary literature

T. Evans, Embedding theorems for multiplicative systems and projective geometries, Proc. American Math. Soc. 3 (1952), 614-620.

Fr 38 R. Frucht, Herstellung von Graphen mit vorgegebener abstrakter Gruppe, Compositio Math. 6 (1938), 239-250.

R. Frucht, Graphs of degree 3 with a given abstract group of automorphisms, Canad. J. Math. 1 (1949), 365-378.

Ha 43 M. Hall, Projective planes, Trans. American Math. Soc. 54 (1943), 229-277.

Z. Hedrlin and E. Mendelsohn, The category of graphs with given subgraph, Canad. J. Math. 21 (1969), 1506-1517.

KS 73 O. H. Kegel and A. Schleiermacher, Amalgams and embeddings of projective planes, Geometriae Dedicata 2 (1973), 379-395.

E. Mendelsohn, Every group is the collineation group of some projective plane, J. Geom. 2 (1972), 97-106.

G. Sabidussi, Graphs with given automorphism group and given graph theoretical properties, Canad. J. Math. 9 (1957), 515-525.

G. Sabidussi, Graphs with given infinite group, Monatsh. Math. 64 (1960), 446-457.

Ti 77 J. Tits, Endliche Spiegelungsgruppen, die als Weylgruppen auftreten, Inventiones Math. 43 (1977), 283-295.

(3) Further results on group universality

L. Babai, Vector representable matroids of given rank with given automorphism group, Discrete Math. 24 (1978), 119-125.

L. Babai, On the abstract group of automorphism, in: Combinatorics (ed. by H. N. V. Temperley), London Math. Soc. Lect. Note Series 52, 1981.

L. Babai and D. Duffus, Dimension and automorphism groups of lattices, Alg. Universalis 12 (1981), 279-289.

O. Bachmann, Embeddings and collineation groups of projective planes, Archiv der Math. 29 (1977), 129-135.

A. Barlotti and K. Strambach, The geometry of binary systems, Advances Math. 49 (1983), 1-105.

G. Birkhoff, Sobre los grupos de automorfismos, Revista Unión Mat. Argentina 11 (1945), 155-157.

E. Fried and J. Kollár, Automorphism groups of fields, in: Universal Algebra (ed. by E. T. Schmidt et al.), Coll. Math. Soc. J. Bolyai 24, 1981.

J. de Groot, Automorphism group of rings, Internat. Congr. Math., Edinburgh 1958.

J. de Groot, Groups represented by homeomorphism groups I, Math. Annalen 138 (1959), 80-102.

Z. Hedrlin and J. Lambek, How comprehensive is the category of semigroups? J. Algebra 11 (1969), 195-212.

E. Mendelsohn, On the groups of automorphisms of Steiner tripel and quadrupel systems, J. Comb. Th. A 25 (1978), 97-104.

E. Mendelsohn, Every (finite) group is the group of automorphisms of a (finite) strongly regular graph, Ars Combinatoria 6 (1978), 75-86.

E. T. Schmidt, Universelle Algebren mit gegebenen Automorphismengruppen und Unteralgebrenverbänden, Acta Sci. Math. Szeged 24 (1963), 251-254.

(4) Further result on homogeneity

G. Higman, B. H. Neumann and H. Neumann, Embedding theorems for groups, J. London Math Soc. 24 (1949), 247-254.

Universal Algebra and its Links with Logic,
Algebra, Combinatorics, and Computer Science
Proc. "25. Arbeitstagung über Allgemeine Algebra", Darmstadt 1983
P. Burmeister et al. (eds.)

183 – 190

WÜNSCHE EINES GEOMETERS AN DIE ALLGEMEINE ALGEBRA

A. Herzer

Obwohl ein großer Teil geometrischer Sachverhalte mit Hilfe von algebraischen Strukturen im weitesten Sinne dargestellt oder erläutert wird, sind die meisten Geometer nicht bereit, Methoden der allgemeinen Algebra (aA) zu benutzen. Um sie von der Nützlichkeit der aA für die Geometrie zu überzeugen, müssen direkt anwendbare Ergebnisse vorliegen; oder Vertreter der aA müssen ihre Resultate schon so weit in die Sprache der Geometrie übersetzt haben, daß ein Geometer einsieht, wie sehr sich Beschäftigung mit aA zugunsten der Geometrie lohnt.
Dieser Beitrag bringt einige Beispiele aus der Sicht des Geometers zusammen mit Fragestellungen, in welchen die aA vielleicht weiterhelfen könnte (oder auch an Grenzen stößt) - zugleich als Einladung an Vertreter der aA, sich mit dem einen oder anderen geometrischen Problem soweit vertraut zu machen, daß Hin- und Rückübersetzung Geometrie -Algebra gelingt und dadurch das Problem der Lösung näher gebracht wird.

A. Die allgemeine Algebra könnte dem Geometer als Sprache/Ordnungsprinzip imponieren, welches ihm den begrifflichen Rahmen für seine algebraischen Hilfskonstruktionen bietet.

1. Bekanntestes Beispiel ist der Begriff des Ternärkörpers zur Koordinatisierung einer projektiven Ebene [18]. Denn dieser läßt sich in der Sprache der aA als Algebra mit genau einer Operation (und diese dreistellig) auffassen, während die Verknüpfungen + und o daraus durch Einsetzen von Konstanten 0 bzw. 1 in die richtigen

Argumente entstehen. Leider bilden die Ternärkörper keine gleichungsdefinierte Klasse; aber gerade dies führt auf das folgende *Problem*: Welches ist die kleinste Varietät, welche die Ternärkörper enthält, und wie lassen sich die zugehörigen Inzidenzstrukturen (gebildet in Analogie zur Konstruktion der projektiven Ebene aus dem Ternärkörper) charakterisieren, sei es allgemein, sei es im endlichen Falle? Entsprechend für größere Varietäten, welche die Ternärkörper enthalten.

2. *Verbandstheorie* (, die wir hier als wichtigen Teil der aA ansehen wollen,) liefert für viele Geometrien eine angemessene Form der Darstellung; so ergibt sich nicht nur die schon klassisch zu nennende verbandstheoretische Charakterisierung der projektiven, Hjelmslev- und allgemeinerer Geometrien [6,4,23], sondern z.B. auch auf dem Gebiet der Matroide (geometrischen Verbände, kombinatorischen Geometrien) die interessante Frage nach der Einbettbarkeit in projektive Geometrien gleichen Ranges, die sich auf semimodulare Verbände ausdehnen läßt [16,24]:
Für einen semimodularen Verband L vom Rang n bezeichnet L_i die Menge der Elemente vom Rang i; für i=1,2,3,n-1 heißen diese bzw. Punkte, Geraden, Ebenen, Hyperebenen. Eine Gerade heißt *Tangente*, wenn sie nur einen Punkt besitzt. Wir nennen einen solchen Verband *lokal projektiv*, wenn [p,1] eine projektive Geometrie ist für alle Punkte p; und L heißt *schlicht*, wenn es eine Abbildung $L_1 \to L_{n-1}$ p→<p> gibt, so daß <p> alle Tangenten durch p enthält.
Eine Möbiusgeometrie läßt sich als schlichter lokal projektiver semimodularer Verband darstellen; ist ein solcher vom Rang >4 (die Möbiusgeometrie vom Rang >2), so ist projektive Einbettung möglich. Die Benz-Ebenen (Möbius-, Laguerre-, Minkowski-Ebenen) lassen sich verbandstheoretisch gemeinsam charakterisieren als semimodulare Verbände vom Rang 4, welche schlicht und lokal projektiv sind [13].
Problem: a) Gibt es außerdem noch weitere geometrische Strukturen, die durch obige Verbände beschreibbar sind, falls man noch verlangt, daß alle Ebenen <p> untereinander isomorph aber keine projektiven Ebenen sind. b) Klassifikation dieser Verbände.
Die immer noch unbeantwortete Frage, ob es überhaupt nichteinbettbare *endliche* Möbiusebenen gibt, führt zur Aufgabe: Man versuche obige Verbände so zu koordinatisieren, daß ihre Ebenen mittels (mehrstelliger) Operationen beschreibbar werden usw. (vgl. in dieser Richtung [9]).

3. Daß für die Beschreibung von Geometrien ein Verband nicht immer ausreicht, zeigt die Arbeit von Faltings [26]: Zur Charakterisierung derjenigen modularen Verbände, welche die Verbände aller Untermoduln eines R-Moduls sind, wird hier noch zusätzlich - entsprechend der Gesamtheit der zyklischen R-Moduln - ein "Punktsystem" benötigt. Dies hat zu einer allgemeineren Konzeption geführt, welche Geometrien adäquater beschreiben soll und diese gegenüber Verbänden aber auch Inzidenzstrukturen qualifiziert. Ein erster Schritt auf diesem Wege ist [20]. *)

4. Während sich Permutationsmengen, R-Moduln gut innerhalb der aA beschreiben lassen, fragt es sich, ob sich auch solche Mischgebilde wie Inzidenzgruppen [15] oder Geometrien versehen mit einer gewissen Gruppe (Menge) von Automorphismen für die Beschreibung mittels aA eignen, also auch Folgerungen für sie aus der aA gezogen werden können, oder ob hier wie vielleicht unter A3 die aA an Grenzen stößt.

B. Die Konstruktion von Geometrien mittels algebraischer Strukturen könnte nach Übersetzung in die Sprache der aA verallgemeinerungsfähig werden und so der Geometrie wieder neues Terrain erschließen. Eine "Geometrie" kann im folgenden - im Gegensatz zu A3 - stets als vollständiger atomistischer nach oben stetiger Verband aufgefaßt werden.

1. Auf einer Tagung über Grundlagen der Geometrie in Utrecht hat R.Baer einmal schematisch die bekannten Methoden dargestellt, wie man aus R-Moduln Geometrien gewinnen kann, [7]:

	projektiv	affin
vollständig	Gesamtheit der zulässigen Untergruppen	Gesamtheit der Restklassen nach zulässigen Untergruppen
Hjemslev	Gesamtheit der zulässigen direkten Summanden	Gesamtheit der Restklassen nach zulässigen direkten Summanden

*) Ebenfalls ein allgemeineres Konzept bietet die Darstellung von Polargeometrien (Quadriken usw.) als Durchschnitts-Halbverbände in [28], für die dann partiell eine Verbindung erklärt wird.

Was entspricht dem in der aA? R.Wille hat in seinen "Kongruenzklassengeometrien" [25] Zusammenhänge zwischen aA und Geometrie untersucht. (Dies Buch liefert eine ausgezeichnete Einführung in die Beziehungen und Fragestellungen zwischen aA und Geometrie und ist in den Auswirkungen seiner Ergebnisse noch längst nicht ausgeschöpft. Ich möchte sowohl Geometern als auch Vertretern der aA seine Lektüre zum ersten Kennenlernen der Gegenseite empfehlen, obwohl es selber natürlich die Dinge von der aA her aufrollt.) Hier wird die (vollständige) affine Geometrie (mit Parallelismus) als Verband der Kongruenzklassen einer Algebra aufgefaßt. Ferner heißt eine Geometrie projektiv koordinatisierbar, wenn sie isomorph zum Verband der Kongruenzrelationen einer Algebra ist. Was in [25] S.51 steht, scheint mir immer noch aktuell: "Es ist sicherlich eine lohnende Aufgabe, die "projektive Koordinatisierung" ausführlich zu untersuchen, zumal man nach dem Satz von Grätzer und Schmidt ... weiß, daß jede Geometrie projektiv koordinatisierbar ist." Nach [25] wäre also im vollständigen Fall die Tabelle von der Form:

projektiv	affin
Kongruenzrelationen	Kongruenzklassen

Problem: Wie läßt sich die "Hjelmslevsche Geometrie" (letzte Zeile der Tabelle von Baer) ebenfalls - stets oder in Spezialfällen - in der Sprache der aA ausdrücken? Untersuchungen in dieser Richtung wären jedenfalls von Interesse.

Ich möchte noch erwähnen, daß inzwischen z.B. die Kongruenzklassengeometrie zu Gruppoiden bereits näher untersucht worden ist; unter nicht allzugroßen Voraussetzungen gelangt man schon zu klassisch bekannten Geometrien [14,17].

2. "Projektiv" wäre ebenfalls die Konstruktion einer Geometrie als Verband der Unteralgebren einer Algebra zu nennen. Hierher gehören die ganzen verdienstvollen Untersuchungen die beim Zusammenhang von Steinertripelsystemen und Squags ihren Ausgang nehmen, z.B. [10,19].

Der Unterschied zwischen den Geometrien in B1 und B2 sieht für eine Gruppe G so aus: In B1 hat man projektiv den Verband der Normalteiler von G und affin den Verband der Nebenklassen der Normalteiler von G; die Geometrie in B2 ist dagegen projektiv der Verband aller Untergruppen von G. In diesem Falle könnte man affin als Geometrie mit Parallelismus den Verband aller Rechtsnebenklassen zu Untergruppen von G wählen.

Problem: Für welche Klassen von Algebren kann man eine geeignete Definition für affine Konstruktionen von Geometrien in Verallgemeinerung obigen Verfahrens für Gruppen gewinnen?

3. Statt in B2 den Verband aller Untergruppen von G zu nehmen kann man auch mit einem nach oben stetigen Mengenverband $\mathcal{M}$ *) von Untergruppen von G arbeiten; das ist Ausgangspunkt von [11]. (Nimmt man für $\mathcal{M}$ speziell eine nichttriviale Partition von Untergruppen von G, so erhält man die Andréschen Translationsstrukturen [1].) Folgendes Beispiel gibt eine leichte Verallgemeinerung [12]: Sei L eine Loop, welche einen Mengenverband $\mathcal{U}$ von Unterloops besitzt, so daß gilt

$$\forall U \in \mathcal{U} \forall u, u' \in U \ \forall a \in L: (u \cdot u') \cdot a = u \cdot (u' \cdot a).$$

Dann sind alle Unterloops aus $\mathcal{U}$ sogar Gruppen. Definiert man affin eine Geometrie $(L,\mathcal{U})_r$, deren Unterräume die Rechtsnebenklassen nach den Unterloops aus $\mathcal{U}$ sind, so ist ein Parallelismus $\parallel$ sinnvoll definierbar durch

$$Ug \parallel U'h \Leftrightarrow U=U' \qquad \forall U,U' \in \mathcal{U} \ \forall g,h \in L.$$

([12] bringt nichttriviale Konstruktionsbeispiele.)

Problem: Welche Klassen von Algebren lassen entsprechende Konstruktionen zu?

4. Arbeiten André's stecken voller Anregungen für die aA; man findet z.B. Fastvektorräume [2] oder schiefaffine Geometrien, z.B. [3]: Die Verbindung zwischen Punkten ist nichtkommutativ. Hier gibt es Zusammenhänge mit Permutationsgruppen und Fastkörpern. In jüngster Zeit ist die Koordinatisierung einer großen Klasse von schiefaffinen Geometrien mittels eines Quaternärkörpers gelungen [22].

5. Sicher gibt es noch mancherlei andere Möglichkeiten, die noch zu finden sind, Algebren und Geometrien einander zuzuordnen. Wünschbar ist dann immer eine interne Charakterisierbarkeit solcher Geometrien. In Verallgemeinerung von A1 könnte man jeder Algebra A mit je einer (m+n-j)-stelligen Operation f_j für $j=0,\ldots,i$ eine Geometrie $\Gamma_{m,n}(A,f_0,\ldots,f_i)$ zuordnen, deren Punktmenge $A^{(m)}$ ist, während ihre

---

*) Das Infimum von Elementen aus $\mathcal{M}$ soll ihr mengentheoretischer Durchschnitt sein.

Hyperebenen von der Form

$$[c,a_1,\dots,a_{n-j}] = \{(x_1,\dots,x_m) \mid f_j(a_1,\dots,a_{n-j},x_1,\dots,x_m)=c\},$$
$$j=0,\dots,i, \qquad c,a_s \in A,\ s=1,\dots,n-j,$$

sind. Mit $f_o(a_1,\dots,a_n,x_1,\dots,x_n) = \sum(x_i-a_i)^2$ usw. kann man z.B. Kugelgeometrien beschreiben, allgemeiner Geometrien, deren Hyperebenen affine algebraische Hyperflächen sind, z.B. affine Darstellungen von Kettengeometrien [27].
Problem: Könnten nicht von daher auch bessere Beziehungen zur algebraischen Geometrie hergestellt werden?

6. Schon länger hat man den Zusammenhang von Geweben und algebraischen Strukturen spezieller Art gesehen [5,18]; vielleicht liegt auch hier ein Ausgangspunkt für die aA? Die Beziehung zwischen algebraischen Eigenschaften von Loops und geometri schen Eigenschaften der zugehörigen 3-Gewebe gewinnt gerade in letzter Zeit wieder neues Interesse [21]. Auf diesem Gebiete wie auf dem Gebiete der zu projektiven Ebenen gehörenden Algebren Quasikörper, Fastkörper Alternativkörper usw. haben Geometer viel Material zusammengetragen und bearbeitet, das gewiß auch von der aA mit Nutzen verwendet werden kann [8].

ZUSAMMENFASSUNG. Ich habe Arbeiten aus dem Gebiet der aA erwähnt, die mir als Geometer etwas sagen, und umgekehrt von einigen Entwicklungen in der Geometrie erzählt, die vielleicht für die aA von Interesse sein könnten, besonders wenn sie auch für die Geometrie ihre Nützlichkeit erweisen will. Zuweilen hat man bisher von den geometrischen Nutzanwendungen der aA noch den Eindruck, daß sie zu sehr auf die Belange der aA zugeschnitten sind und daher den genuinen Geometer nicht so recht zu interessieren vermögen. Man müßte noch mehr die Grenze zur Geometrie überschreiten, um in Konfrontation mit deren ureigensten Problemen neue Impulse zur Weiterentwicklung der aA in Richtung auf deren Bewältigung zu erhalten.

BIBLIOGRAPHIE

1 J.André, Über Parallelstrukturen, Teil II: Translationsstrukturen. Math.Z.76(1961)155-163

2 ---, Lineare Algebra über Fastkörpern. Math.Z.136(1974)295-313

3 ---, On finite non-commutative affine spaces. In: Combinatorics (Breukelen 1974), ed. by M.Hall jr. and J.H. van Lint, 2nd ed., pp. 65-113. Amsterdam: Mathematical Centre 1975

4 B.Artmann, Uniforme Hjelmslev-Ebenen und modulare Verbände. Math.Z.111(1969)15-45

5 R.Artzy, Eigenschaften von Viergeweben in allgemeiner Lage. Math.Ann.126(1953)336-342

6 R.Baer, Linear Algebra and Projective Geometry, New York 1952

7 ---, Hjelmslevsche Geometrie. Algebraical and Topological Foundations of Geometry, Proceedings of a Colloquium, Utrecht, 1-4, Oxford, London, New York, Paris, 1962

8 R.H.Bruck, A Survey of Binary Systems, Springer Berlin Heidelberg New York, 1966

9 G.Ewald, Kennzeichnungen der projektiven dreidimensionalen Räume und nichtdesarguessche räumliche Strukturen über beliebigen Ternärkörpern. Math.Z.75(1961)395-418

10 B.Ganter und R.Metz, Kombinatorische Algebra:Koordinatisierung von Blockplänen. In: Beiträge zur Geometrischen Algebra,Hrg. H.J.Arnold, W.Benz und H.Wefelscheid, 1977, pp.111-124, (Birkhäuser Basel Stuttgart)

11 A.Herzer, Halbprojektive Translationsgeometrien. Mitt.Math.Sem. Gießen, Heft 127(1977)

12 ---, Translationsstrukturen, die weder axial noch zentral sind. Geom.Ded.8(1979)163-178

13 ---, Semimodular Locally Projective Lattices osf Rank 4 from v.Staudt's Point of View. In: Geometry - von Staudt's Point of View, Hrsg. P.Plaumann und K.Strambach,1981, pp.374-400, (D.Reidel, Dordrecht Boston London)

14 Th.Ihringer, On Grupoids Having a Linear Congruence Class Geometry. Preprint, Darmstadt 1981.

15 H.Karzel, Zweiseitige Inzidenzgruppen. Abh.Math.Sem.Univ.Hamburg 29(1965)118-136

16 H.Mäurer, Ein axiomatischer Aufbau der mindestens 3-dimensionalen Möbiusgeometrie. Math.Z.103(1968)282-305

17 A.Pasini, On the Finite Transitive Incidence Algebras. Boll. Un.Mat.Ital.(5) 17-B(1980)373-389

18 G.Pickert, Projektive Ebenen. Springer Berlin Göttingen Heidelberg 1955

19 R.W.Quackenbush, Algebraic aspects of Steiner quadruple systems. Proceedings of the conference on algebraic aspects of combinatorics, University of Toronto 1975, pp.265-268

20 Stefan Schmidt, Staatsexamensarbeit Hamburg 1982: "Darstellung von Inzidenzräumen und Hüllensystemen durch Verbände."

21 K.Strambach, Geometry and Loops. In: Geometry and Groups. Lecture Notes in Mathematics 893, Springer Berlin Heidelberg New York

22 H.Tecklenburg, Zur algebraischen Darstellung fastaffiner Räume. Vortrag Oberwolfach, Okt.1982

23 R.Wille, Verbandstheoretische Charakterisierung n-stufiger Geometrien. Arch.Math.18(1967)465-468

24 ---,On Incidence Geometries of Grade n. Atti Conv.Geom.Comb. Appl.Perugia 1971, pp.421-426

25 ---,Kongruenzklassengeometrien. Lecture Notes in Mathematics 113, Springer Berlin Heidelberg New York 1970

26 K.Faltings, Modulare Verbände mit Punktsystem. Geom.Dedic.4 (1975)105-137

27 W.Benz, Vorlesungen über Geometrie der Algebren. Springer Berlin Heidelberg New York 1973

28 F.D.Veldkamp, Polar Geometry, III.Axioms for Polar Geometry. Indag.Math.21(1959)527-533.

Universal Algebra and its Links with Logic,
Algebra, Combinatorics, and Computer Science
Proc. "25. Arbeitstagung über Allgemeine Algebra", Darmstadt 1983
P. Burmeister et al. (eds.)

191 – 203

AN ALGEBRAIC DECOMPOSITION THEORY FOR DISCRETE STRUCTURES

R. H. Möhring

ABSTRACT: In this paper we give an outline of a general algebraic decomposition theory for discrete structures. This theory unifies and generalizes the so-called substitution decomposition or modular decomposition of e.g. Boolean functions, set systems or relations, which has many applications in discrete mathematics, operations research and computer science. Major results in this theory are a general Jordan-Hölder theorem and a Church-Rosser property for composition series of these structures, properties of the congruence partition lattices, and characterization theorems for highly decomposable structures.

1. INTRODUCTION AND BACKGROUND:

This paper deals with a general decomposition theory for discrete structures which unifies and generalizes decomposition approaches to structures such as Boolean functions, set systems or relations. These approaches occur in the literature unrelatedly under different names such as substitution decomposition, modular decomposition, lexicographic decomposition or X-join. They have a large scope of applications in discrete mathematics, operations research and computer science, ranging from switching design (disjunctive decomposition of Boolean functions [1],[10]) via automata theory (decomposition of the control relation of an automaton [21]), decision theory (decomposition of utility functions [12]), reliability theory (modular decomposition of coherent systems [2],[5]) and game theory (decomposition of simple n-person games [15],[20]) to combinatorial (optimization) problems over graphs, networks and independence systems or clutters, cf [6], [17] for an overview. For these optimization problems, the objective can be obtained in a two-step procedure by exploiting a (given) decomposition in a natural way. Furthermore, for many such problems it can be shown that the substituion decomposition is in fact the only possible two-step decomposition under certain, rather weak assumptions on the decomposition approach, cf [17] for more details.

For all these structures considered above, the decomposition can be seen as the inverse of a composition or substitution operation which works as follows:

Let S be a class of structures (e.g. graphs, set systems etc.). Let S' be a structure from S on a set A' and let, for each $\beta \in A'$, S_β be a structure from S on a set A_β, where the sets A_β are non-empty and pairwise disjoint. The composition operation assigns to S' and S_β, $\beta \in A'$, a unique structure S on $A := \bigcup_{\beta \in A} A_\beta$, which contains the structures S_β, $\beta \in A'$, as substructures (i.e. S_β equals the restriction $S|A_\beta$ of S to A_β), and in which the relationship between the different S_β is defined via S'. This structure S is called the composition of the structures S' and S_β, $\beta \in A'$, and is denoted by $S = S'[S_\beta, \beta \in A']$ or $S'[S_1,\dots,S_m]$ if $A' = \{1,\dots,m\}$.

For k-ary ($k \geq 2$) relations R' on A', R_β on A_β, $\beta \in A'$, the composition $R = R'[R_\beta, \beta \in A']$ is defined by

$$(\alpha_1,\dots,\alpha_k) \in R :\Longleftrightarrow \begin{cases} (\alpha_1,\dots,\alpha_k) \in R_\beta \text{ for some } \beta \in A' \text{ or} \\ (\alpha_1,\dots,\alpha_k) \in A_{\beta_1} \times \dots \times A_{\beta_k} \text{ for some } (\beta_1,\dots,\beta_k) \in R', \text{ with } |\{\beta_1,\dots,\beta_k\}| > 1. \end{cases}$$

Special cases of this composition are the X-join for undirected graphs [13], the lexicographic sum for partial orders [22], and the modular decomposition of networks [2]. An example for graphs is given in Figure 1.

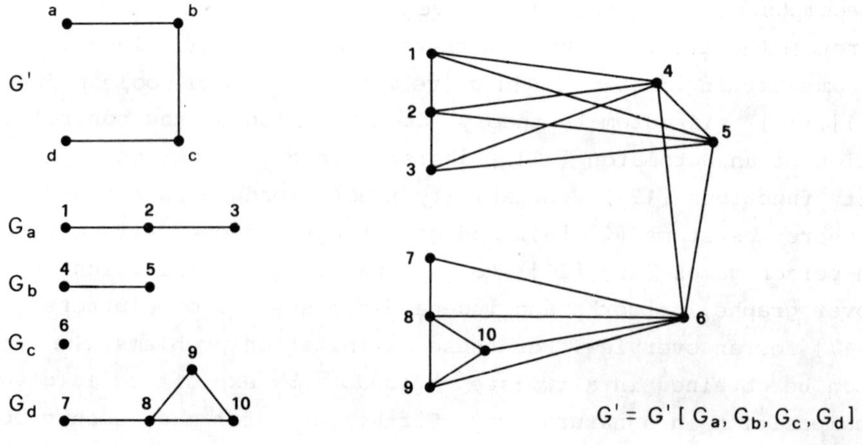

Figure 1: A composition of graphs

For <u>set systems</u> $\mathcal{T}'$ on A', $\mathcal{T}_\beta$ on A_β, $\beta \in A'$, the decomposition $\mathcal{T} = \mathcal{T}'[\mathcal{T}_\beta, \beta \in A']$ is defined by

$$T \in \mathcal{T} :\Longleftrightarrow \begin{cases} \text{there exist } T' \in \mathcal{T}' \text{ and } T_\beta \in \mathcal{T}_\beta \text{ for each } \beta \in T' \\ \text{such that } T = \bigcup_{\beta \in T'} T_\beta . \end{cases}$$

Special cases here are the composition for clutters (i.e. set systems of pairwise incomparable sets) [4], and the modular decomposition of coherent systems [2][5].

For an illustration, consider the clutters $C' = \{\{a,b,\}, \{b,c\}, \{c,d\}\}$, $C_a = \{\{1,2\},\{2,3\}\}$, $C_b = \{\{4,5\}\}$, $C_c = \{\{6\}\}$ and $C_d = \{\{7,8\}, \{8,9,10\}\}$ on the sets $\{a,b,c,d\}$, $\{1,2,3\}$, $\{4,5\}$, $\{6\}$, $\{7,8,9,10\}$, respectively. Then $C = C'[C_a, C_b, C_c, C_d] = \{\{1,2,4,5\},\{2,3,4,5\}, \{4,5,6\},\{6,7,8\},\{6,8,9,10\}\}$.

For Boolean functions $F'(y_1,\ldots,y_m)$, $F_1(x_1,\ldots,x_{k_m}),\ldots, F_m(x_{k_{m-1}+1},\ldots,x_n)$ with disjoint sets of variables $\{x_1,\ldots,x_{k_1}\}$, $\{x_{k_1+1},\ldots,x_{k_2}\},\ldots,\{x_{k_{m-1}+1},\ldots,x_n\}$, the composition $F = F'[F_1,\ldots, F_m]$ is defined by

$$F(x_1,\ldots,x_n) = F'(F_1(x_1,\ldots,x_{k_1}),\ldots,F_m(x_{k_{m-1}+1},\ldots,x_n)),$$

i.e. it corresponds to the notion of <u>disjunctive decomposition</u> in switching theory [1][10].

As an example, let $F'(y_1,\ldots,y_4) = y_1y_2 + y_2y_3 + y_3y_4$, $F_1(x_1,x_2,x_3) = x_1x_2 + x_2x_3$, $F_2(x_4,x_5) = x_4x_5$, $F_3(x_6) = x_6$, and $F_4(x_7,\ldots,x_{10}) = x_7x_8 + x_8x_9x_{10}$. Then $F = F'[F_1,\ldots,F_4]$ is given as

$$\begin{aligned} F(x_1,\ldots,x_{10}) &= (x_1x_2 + x_2x_3)(x_4x_5) + (x_4x_5)x_6 + x_6(x_7x_8 + x_8x_9x_{10}) \\ &= x_1x_2x_4x_5 + x_2x_3x_4x_5 + x_4x_5x_6 + x_6x_7x_8 + x_6x_8x_9x_{10} \end{aligned}$$

In all cases, we also say that the composed structure $S = S'[S_\beta, \beta \in A']$ is obtained by <u>substituion</u> of the elements $\beta \in A'$ in S by the structures S_β, $\beta \in A'$.

The composition is <u>proper</u> if $|A'| > 1$ and $|A_\beta| > 1$ for some $\beta \in A'$. A structure S on A is said to be <u>decomposable</u> if it has a representation as a proper composition $S = S'[S_\beta, \beta \in A']$. In that case, the associated partition $\pi = \{A_\beta \mid \beta \in A'\}$ of A into the ground sets A_β of S_β is called a (non-trivial) <u>congruence partition</u> of S, the structures S_β are called the <u>substructures</u> associated with π and the structure S' is called the <u>quotient</u> of S modulo π and is denoted by S/π. A structure S which is not decomposable is called <u>indecomposable</u> or <u>prime</u>.

Though the composition operations for the three classes of structures are defined in quite different ways, there are some very strong links between them.

The link between relations and set systems is given by graphs and conformal clutters (i.e. clutters of maximal cliques of a graph), cf [7]. If we denote by $\mathcal{C}(G)$ the clutter of maximal cliques of graph G, and by $G(\mathcal{C})$ the graph defining the conformal clutter $\mathcal{C}$, then

$$\mathcal{C}(G'[G_\beta,\beta \in A']) = \mathcal{C}(G')[\mathcal{C}(G_\beta),\beta \in A'] \text{ and}$$
$$G(\mathcal{C}'[\mathcal{C}_\beta,\beta \in A']) = G(\mathcal{C}')[G(\mathcal{C}_\beta),\beta \in A'],$$

i.e., the composition operations are essentially equal.

The link between set systems and Boolean functions is given by monotonic Boolean functions and the corresponding clutters of their prime implicants, cf [3]. If we denote by $\mathcal{C}(F)$ the clutter of prime implicants of F, and by $F(\mathcal{C})$ the monotonic Boolean function whose prime implicants are the sets of $\mathcal{C}$, then

$$\mathcal{C}(F'[F_1,\ldots,F_m]) = \mathcal{C}(F')[\mathcal{C}(F_1),\ldots,\mathcal{C}(F_m]) \text{ and}$$
$$F(\mathcal{C}[\mathcal{C}_1,\ldots,\mathcal{C}_m]) = F(\mathcal{C})[F(\mathcal{C}_1),\ldots,F(\mathcal{C}_m)].$$

Furthermore, if $F = F'[F_1,\ldots,F_m]$ where F is monotonic but F' and the F_i need not be, then there exist monotonic Boolean func- $G',G_1,\ldots,G_m$ with the same variables as $F',F_1,\ldots,F_m$, respectively, such that $F = G'[G_1,\ldots,G_m]$. In other words, the decomposition possibilities of a monotonic Boolean function are the same as those for its clutter of prime implicants.

These links are also demonstrated by the above examples, where $\mathcal{C} = \mathcal{C}(G)$ and $F = F(\mathcal{C})$.

These links, together with the large scope of applications, formed the reasons for the introduction and investigation of the general model presented in this paper. A detailed treatment of this model is given in [16], cf also [17].

2. THE ALGEBRAIC MODEL

In the general decomposition model we consider a "concrete" category $\mathcal{K}$, whose objects are called structures and are denoted by S, T etc. "Concrete" means that each structure is defined on an underlying set $A = A_S$ (the base set of S), and that each homomorphism (or morphism in categorial terminology) from S to T is a mapping from A_S into A_T.

A special role will be played by the surjecitve and injective homomorphisms which (in accordance with the usual algebraic terminology, but different from categorial terminology) will be referred to as epimorphisms and monomorphisms, respectively. For structures S,T from K, let Hom(S,T), Epi(S,T), and Mono(S,T) denote the sets of homomorphisms, epimorphisms, and monomorphisms from S to T, respectively. Let S denote the class of structures in K. Two structures S and T are isomorphic if there exists a bijective mapping f such that $f \in$ Hom(S,T) and $f^{-1} \in$ Hom(T,S).

In addition to the usual categorial properties, we impose two groups of conditions, (M1) - (M5) and (M6) - (M8). The first group provides us with elementary algebraic properties needed to define quotients and substructures, while the second group deals with the relationship between these notions. It should already be noted that (M1) - (M6) hold in the familiar algebraic theories (e.g. the theory of groups, rings, etc.) and that it is, in fact, only conditions (M7) and M8) which are "non-algebraic". These two conditions may thus be viewed as representing the special character of the substitution-decomposition.

(M1) Each $f \in$ Hom(S,T) has an epi-mono-factorization, i.e., there exist $U \in S$, $g \in$ Epi(S,U), $h \in$ Mono(U,T) such that $f = h \circ g$.

(M2) Structure is abstract, i.e., given a structure S on A and a bijection $f: A \to B$, there exists a unique structure T on B such that $f \in$ Hom(S,T) and $f^{-1} \in$ Hom(T,S).

(M3) Given a structure S on A and a surjection f from A onto a singleton A_o, there exists a structure S_o on A_o such that $f \in$ Epi(S,S_o).

(M4) If $h \in$ Epi$(S,T_1) \cap$ Epi(S,T_2) then $T_1 = T_2$.

(M5) If $g \in$ Mono$(S_1,T) \cap$ Mono(S_2,T) then $S_1 = S_2$.

We call T a quotient of S if there is a partition π of A_S such that the associated natural mapping $\eta_\pi \in$ Epi(S,T). Then π is called a congruence partition of S, and the uniquely determined (because of (M4)) quotient T is denoted by S/π. V(S) denotes the system of congruence partitions of S.

S is called prime or indecomposable if $V(S)$ contains no proper congruence partition, i.e. if $C \in \pi \in V(S)$ implies $C = A$ or $|C| = 1$.

T is called a substructure of S, if there exists a subset B of A_S such that the associated inclusion mapping $inc_B^A \in Mono(T,S)$. In this case, B is called an S-autonomous set, and the uniquely determined (because of (M5)) substructure T is denoted by $S|B$. $A(S)$ denotes the system of S-autonomous sets.

One obtains by standard arguments that epimorphisms and monomorphisms correspond essentially to quotient structures and substructures, which in turn can for a given structure S be "internally" described by its system of congruence partitions $V(S)$ and its system of S-autonomous sets $A(S)$.

The question then is how to embed the substitution decomposition into this general framework. This is not quite obvious, since, in general, there is no "natural" notion of homomorphism. There are, however, natural notions of "quotient" and "substructure" which may be used to define homomorphisms appropriately. To this end, let S be a structure (i.e. a relation, set system of Boolean function) obtained by substituion, i.e. $S = S'[S_\beta, \beta \in A']$, where S' is a structure on A' and, for each $\beta \in A'$, S_β is a structure on A_β. Then the associated surjective mapping $h : A \to A'$ with $h(\alpha) = \beta$ iff $\alpha \in A_\beta$ is used to define the epimorphisms in the general model, while the associated inclusion mappings $inc_{A_\beta}^A$ are used to define the monomorphisms.

For graphs G, G', for instance, the surjective homomorphisms h from G onto G' are given by the condition that all nodes α, β with $h(\alpha) \neq h(\beta)$ are adjacent in G iff $h(\alpha)$ and $h(\beta)$ are adjacent in G'.

Arbitrary homomorphisms are then defined as the composition of finitely many epimorphisms and monomorphisms. The verification of (M1) - (M5) is somewhat laborious. However, the obtained interpretation of the substitution decomposition in terms of an algebraic homomorphism theory permits the application of methods and concepts from universal algebra. Furthermore, there are examples of the general model which cannot be obtained from a substitution operation, i.e. examples in which it is not possible to "uniquely reconstruct" a structure S from a quotient S/π and the substructures $S|B$, $B \in \pi$ (cf [16]).

The second group of conditions of the algebraic model then reflect possiblities for constructing new congruence partitions from already known ones. They are motivated by the properties of the substitution operation.

(M6) $\pi, \sigma \in V(S)$ and $\pi \leqslant \sigma \Rightarrow \mathrm{Epi}(S/\pi, S/\sigma) \neq \emptyset$. (Induced Homomorphism Theorem)

(M7) (i) $\pi \in V(S) \Rightarrow B \in A(S)$ for all $B \in \pi$.

(ii) If $\pi = \{L_1, \ldots, L_n, \{\alpha\} \mid \alpha \in A \setminus \bigcup_{i=1}^{n} L_i\}$ is a partition of A_S with $L_i \in A(S)$, $i = 1, \ldots, n$, then $\pi \in V(S)$.

(connection between congruence partitions and autonomous sets)

(M8) $\pi = \{B_i \mid i \in I\} \in V(S)$, $\tau = \{C_j \mid j \in J\} \in V(S|B_{i_o})$ for some $B_{i_o} \in \pi \Rightarrow \sigma = \{B_i, C_j \mid i \in I \setminus \{i_o\}, j \in J\} \in V(S)$.

(refinement property of congruence partitions w.r.t. substructures)

These are all the conditions of the general model. They are chosen as weak as possible in order to cover <u>all</u> the known applications (also in the infinite case) but still strong enough to obtain structural insights. In some cases (e.g. relations) even stronger properties hold, cf (M7)*, (M9), (M10) below.

3. THE SYSTEM OF CONGRUENCE PARTITIONS

We shall now consider the system $V(S)$ of congruence partitions of a structure S, viewed as a suborder of the partition lattice $Z(A)$ of the base set A of S, where the ordering relation is the refinement relation $\leqslant$ for partitions. Of course, $V(S)$ always contains $\pi^o = \{\{\alpha\} \mid \alpha \in A\}$ and $\pi^1 = \{A\}$ because of (M2) and (M3).

In general, $V(S)$ need not be a lattice (e.g. for infinite clutters, cf [16][17]). We have, however, the following theorem:

THEOREM 3.1: $V(S)$ contains a maximal chain of finite length iff it is finite. In that case, $V(S)$ is an upper semimodular sublattice of the partition lattice $Z(A)$.

This result will be essential for the Jordan-Hölder theorem in the next section.

Under stronger properties (which hold e.g. for relations and certain classes of set systems), additional lattice theoretical results can be derived. These properties are as follows:

<u>Infinite construction in $V(S)$</u>

(M7)* $\pi \in V(S)$ iff $L \in A(S)$ for all $L \in \pi$

<u>Additional properties of $A(S)$</u>

(M9) If $C_1, C_2 \in A(S)$ overlap, then $C_1 \setminus C_2 \in A(S)$ and $C_2 \setminus C_1 \in A(S)$

(M10) If $(C_i)_{i \in I} \subseteq A(S)$ and $\bigcap_{i \in I} C_i \neq \emptyset$, then $\bigcap_{i \in I} C_i \in A(S)$ and $\bigcup_{i \in I} C_i \in A(S)$

In the general model, (M7)* and (M9) may not hold, while (M10) is only valid for finite I.

We first characterize <u>complemented</u> congruence partition lattices. It turns out that only certain "highly decomposable" structures, viz. degenerate and linear structures have (non-trivial) complemented congruence partition lattices.

Call a structure S <u>degenerate</u> if each non-empty subset of A_S belongs to $A(S)$, and <u>linear</u> if there is a linear ordering $\leqslant$ on A_S such that $A(S)$ consists of all the $\leqslant$-convex sets (i.e. all sets B for which $\alpha, \beta \in B$, $\alpha \leqslant \gamma \leqslant \beta$ implies that $\gamma \in B$).

THEOREM 3.2: Under the above assumptions $V(S)$ is complemented iff one of the following conditions holds:

1. S is prime. Then $V(S)$ is a 2-element Boolean algebra.
2. S is linear and the associated linear order $\leqslant$ is locally finite (i.e. each interval of $\leqslant$ is finite). Then $V(S)$ is a Boolean algebra.
3. S is degenerate. Then $V(S)$ is the partition lattice $Z(A)$.

2 is a special case of a theorem of Hashimoto [14, Th. 8.4] stating that the congruence partition lattice of a distributive lattice (considered as an algebra w.r.t. the join and meet operations) is a Boolean algebra iff the lattice is locally finite.

This follows from the fact that linear orders are lattices, and that the homomorphisms defined via the substitution decomposition coincide for this case with the lattice homomorphisms.

For the characterization of modular and distributive congruence partition lattices we need to specify the degree to which a structure is degenerate.

Let S be a structure on A and let Deg(S) be the set of all degenerate structures which are a substructure of some homomorphic image of S. Then the maximum cardinality of a base set of the structures of Deg(S) (or ∞, if the maximum does not exists) is called the degeneration degree of S and is denoted by deg(S).

THEOREM 3.3: Under the above assumptions, $\mathcal{V}(S)$ is modular (distributive) iff $\deg(S) \leq 3$ (≤ 2).

The proof shows that $\deg(S) = n$ implies that $\mathcal{V}(S)$ contains an interval isomorphic to the congruence partition lattice of a degenerate structure on n elements, i.e. isomorphic to the partition lattice $\mathcal{Z}(n)$ of $\{1,\ldots n\}$. This gives the easy direction of the proof. In the other direction one shows that $\deg(S) \leq 3$ (2) implies that complements in $\mathcal{V}(S)$ are incomparable (unique). Because of the representation of any interval $[\pi,\sigma] \subseteq \mathcal{V}(S)$ as the direct product

$$[\pi,\sigma] = \bigtimes_{B\in\sigma/\pi} \mathcal{V}((S/\pi)\,|\,B)$$ (where σ/π is the partition of S/π induced by σ)

also relative complements in $\mathcal{V}(S)$ are then incomparable (unique).

Altogether, these results show that lattice theoretical properties strongly depend on the properties of degenerate and linear structures. For relations, set systems and Boolean functions, these highly decomposable structures can be completely characterized, thus leading to stronger structural insights.

THEOREM 3.4:

a) A k-ary relation ($k \geq 2$) R on A with $|A| \geq 3$ is degenerate iff, up to reflexive tuples $(\alpha,\ldots,\alpha)$, $R = \emptyset$ or $R = A^k$. It is linear iff it is a linear orderung up to reflexive pairs (α,α).

b) A set system T on A (which covers A) is degenerate iff one of the following conditions applies:

1. $T = \{\{\alpha\} \mid \alpha \in A\}$.
2. $T \cup \{\emptyset\}$ is an ideal of the power set of $P(A)$ of A.
3. There exist an ideal I and a proper dual ideal F of $P(A)$ such that $T = I \cap F$. ($I = P(A)$ is possible)

There are no linear set systems.

c) A Boolean function $F(x_1,\dots,x_n)$ without inessential variables is degenerate iff $F(x_1,\dots,x_n) = x_1^{\varepsilon_1} * \dots * x_n^{\varepsilon_n}$, where $x_j^{\varepsilon_j}$ denotes x_j or $\overline{x}_j$ and $*$ stands for Boolean addition (+), Boolean multiplication ($\cdot$) or ring sum addition ($\oplus$).
There are no linear Boolean functions.

4. COMPOSITION SERIES

We shall now consider the successive factorization of a structure by means of homomorphisms.

Let S be a structure on A. A composition series of S is a maximal finite sequence $S = S_o, S_1,\dots,S_n$ of pair-wise non-isomorphic structures S_i on A_i with $|A_n| = 1$ such that there exists an epimorphism $h_i \in Epi(S_{i-1}, S_i)$ for each $i = 1,\dots,n$.

Since the sequence is supposed to be of maximal length, the congruence partition $\pi_i \in V(S_{i-1})$ induced by h_i is an atom in $V(S_{i-1})$. It then follows that h_i maps exactly one non-trivial, prime substructure $S_{i-1}|B_i$ of S_{i-1} onto one element of S , and maps the elements not in B_i bijectively. The prime substructures $S_o|B_1, S_1|B_2,\dots, S_{n-1}|B_n = S_{n-1}$ are called the factors of the composition series $S = S_o, S_1,\dots,S_n$.

It is easily seen that composition series essentially correspond to the maximal chains in $V(S)$. So, because of Theorem 3.1, S has a composition series iff $V(S)$ is finite. Making use of the upper-semimodularity of $V(S)$ and in particular (M7) and (M8) one obtains the following Jordan-Hölder-type theorem for composition series of structures fulfilling (M1) - (M8).

THEOREM 4.1: Any two composition series of S have the same length and the same factors up to isomorphism and rearrangement.

It should be noted that this theorem does not follow from general "algebraic" Jordan-Hölder theorems as in [11][18]. These are usually based on the commutativity of the congruence relations of the algebra or certain generalizations thereof. Such a property does not hold for the structures considered here. This becomes, for instance, apparent by the fact that the congruence partition lattices are (in the finite case) only upper semimodular, whereas for algebras with commuting congruence relations they are modular. In this respect, the structures considered here constitute a separate case, in which a Jordan-Hölder Theorem holds under conditions different from the usual ones.

A further indication as to the different nature of this result is an "Church-Rosser" property [19] for composition series which is known for relations and set systems, and which does not hold for algebras, in general: Any two composition series of S have the same last factor up to isomorphism.

In the general model, this property does not follow from (M1) - (M8), but requires an additional assumption, which holds in the special cases of Section 1.

(M11) Let $S|B$ and $S|C$ be prime substructures of S such that $B \cap C \neq \emptyset$. Then $S|B$ and $S|C$ are isomorphic.

THEOREM 4.2: If the category K fulfills (M1) - (M8) and (M11), then any two composition series of a structure S have the same last factor up to isomorphism.

CONCLUDING REMARKS: We have demonstrated that the substitution decomposition of discrete structures can be embedded into a general algebraic decomposition theory which has nice structural properties such as a Jordan-Hölder theorem and, under additional assumption, a Church-Rosser property. The general results yield, in turn, new insights for the decomposition of special structures such as relations, set systems or Boolean algebras.

There are of course more questions concerned with the substitution decomposition than could be treated in this paper. One such question regards the <u>algorithmic complexity</u> of the decomposition, a vital question w.r.t. applications. Here the Church-Rosser property and the characterization results in Theorem 3.4 lead to a representation of $A(S)$ into a tree (the so-called composition tree of S) which serves as a data structure for decomposition algorithms, cf [6][16].

Another question concerns related decomposition models such as the split decomposition [8],[9], which (in the finite case) may be considered as a symmetric generalization of the substitution decomposition. This generalization also leads to decomposition trees and stronger characterization theorems for highly decomposable structures in the sense of Theorem 3.4, but does not seem to lend itself for an algebraic treatment, cf [17].

REFERENCES

[1] ASHENHURST, R.L.: The decomposition of switching functions, in: Proceedings of the international symposium on the theory of switching (Part I), Harvard University Press, Cambridge, 1959

[2] BARLOW, R.E., PROSCHAN, F.: Statistical Theory of Reliability and Life Testing, Holt, Rinehart and Winston, New York, 1975

[3] BILLERA, L.J.: Clutter decomposition and monotonic Boolean functions, Annals of the New York Academy of Sciences 175 (1970), 41-48

[4] BILLERA, L.J.: On the composition and decomposition of clutters, J.Comb.Th. B 11 (1971), 234-245

[5] BIRNBAUM, Z.W., ESARY, J.D.: Modules of coherent binary systems, SIAM J. Applied Math. 13 (1965), 444-462

[6] BUER, H., MÖHRING, R.H.: A fast algorithm for the decomposition of graphs and posets, to appear in Math. Oper. Res., extended abstract in Methods of Oper. Res. 37 (1980), 259-264

[7] CHEIN, M., HABIB, M., MAURER, M.C.: Partitive Hypergraphs, Discr. Math. 37 (1981), 35-50

[8] CUNNINGHAM, W.H.: Decomposition of directed graphs, SIAM J. Algebraic and Discrete Structures 3 (1982), 214-228

[9] CUNNINGHAM, W.H., EDMONDS, J.: A combinatorial decomposition theory, Can. J. Math. 32 (1980), 734-765

[10] CURTIS, H.A.: A New Approach to the Design of Switching Circuits, Van Nostrand, Princeton, 1962

[11] GOLDIE, A.W.: The scope of the Jordan-Jölder theorem in abstract algebra, Proc. London Math. Soc. 2 (1952), 349-368

[12] GORMAN, W.M.: The structure of utility functions, Rev. Econ. Studies 35 (1968), 367-390

[13] HABIB, M., MAURER, M.C.: On the X-join decomposition for undirected graphs, J. Appl. Discr. Math. 3 (1979), 198-207

[14] HASHIMOTO, J.: Ideal theory for lattices, Math. Japanicae 2 (1952), 149-186

[15] MEGIDDO, N.: Tensor decomposition of cooperative games, SIAM J. Appl. Math. 29 (1975), 388-405

[16] MÖHRING, R.H.: Dekomposition diskreter Strukturen mit Anwendungen in der kombinatorischen Optimierung, Habilitationsschrift, Techn. Univ. of Aachen, 1982

[17] MÖHRING, R.H., RADERMACHER, F.J.: Substitution decomposition of discrete structures and connections with combinatorial optimization, to appear in Annals of Discrete Mathematics (1983/84)

[18] RICHTER, G.: Kategorielle Algebra, Akademie Verlag, Berlin, 1979

[19] SETHI, R.: Testing for the Church-Rosser property, J. ACM 21 (1974), 671-679

[20] SHAPLEY, L.S.: On Committees, in: New Methods of Thought and Procedure, F. Zwicky, A.G. Wilson (eds.), Springer Verlag, Berlin, New York (1967), 246-270

[21] STRASSNER, K.: Zur Strukturtheorie endlicher nichtdeterministischer Automaten I. Zum Verband der 1-Kongruenzen von endlichen Relationalsystemen, Elektronische Informationsverarbeitung und Kybernetik 17 (1981), 113-120

[22] WILLE, R.: Lexicographic decomposition of ordered sets (graphs), Preprint 705, Fachbereich Mathematik, Techn. Univ. of Darmstadt, 1983

Rolf H. Möhring
Lehrstuhl für Informatik IV
Technical University of Aachen
Templergraben 64
5100 Aachen
West Germany

Universal Algebra and its Links with Logic,
Algebra, Combinatorics, and Computer Science
Proc. "25. Arbeitstagung über Allgemeine Algebra", Darmstadt 1983
P. Burmeister et al. (eds.)

204 – 215

IMPORTANCE OF UNIVERSAL ALGEBRA FOR COMPUTER SCIENCE

H. Andréka, I. Németi

Within the technological - practical branch of our present technical - scientific culture it is exactly the field of computer engineering whose development took a completely new, unusual and very exciting turn. We face here a problem whose handling requires new ways and schemes of thinking and also new conceptual systems. Moreover, it seems necessary that these new schemes of thinking and principles stem from and be worked out in the field of theoretical mathematics. To give a feeling of the qualitative change (called qualitative jump) of the situation after the advent of computers and cybernetics, we mention the following: while the industrial tools and machines produced earlier were aimed at the physical and energetical support of men, the new product - the programmed computer - is to support men's intelligence [37,4],[2]§14/7 . This latter industrial product, the programmed computer, is an amazingly "intelligent" system, able to perform tasks which require complex brain work, even if this performing requires intelligence [37]. This machine is especially well suited to be applied in those spheres of activities people generally do not like to work in because these activities do not require enough creativity, do not help to develop, unfold or enrich their personalities. According to our present knowledge, in our industrial society it is possible, <u>in principle</u>, to leave all these quite uninteresting spheres of activities completely to computers. (Certainly this change would not decrease the number of work possibilities, but quite in contrary increase it, since in a society organized in such a complex way there is an immense need of creative work. As to examplify this we recall that against all expectations the number of work possibilities grew when the transformation of chemical energy into kinetical energy was taken over from men by engines.) Though, as mentioned, it is possible, in principle, to mechanize all noncreative activities, when trying to reach this aim the specialists ran into obstacles which made impossible (at least at the moment) not only this but also other more down to earth applications already of vital importance. In short: it is possible in principle and impossible in practice. The above obstacles are not some recognized, unavoidable, unsurmountable laws of nature, but quite the contrary - lack of knowledge. The obstacle is nothing but the position occupied by computer engineering in the structure of technical - scientific culture in our days; at this point there are essential defects and these form the obstacles. Let us look into the matter in more detail.

The final industrial product, the programmed computer, consists of two main parts: the physical machine and the programs. The first part, i.e. the tangible physical machine is called hardware (while the second one is called software). The hardware as such is amazingly "dull", it is not suited for performing almost any task. The hardware is made by electrical engineers, hence the manufacture technology of products of the electrical industry can more or less support its production. It is the technology of software production which causes trouble. The programs belonging to the machine are extremely important, they form the machine's "soul" and whole knowledge [37]. The key problem of today's computer engineering is connected to the production of programs. This problem is so burning that it was even named software crisis. Thus the problem was caused by the appearance of a new industrial product, the computer program, in our technical - scientific culture. In fact, developed countries spend more and more of their budget on the production of programs. The manufacturing of programs should be realized on an industrial scale and a production technology for programs should be elaborated. With other more traditional products (e.g. cars, engines) in similar situations in our technical - scientific culture one can observe the recurrence of a very typical scheme. A manufacture technology for the product comes into being having as a background a whole hierarchy of subcultures. Let us take for example the case of electrical machines. The technology of electrical machines is based on electrical engineering, which is based on theoretical electrical engineering, which is based on theoretical electricity, which is based on theoretical physics (or can be conceived of as a part of it). (This list is claimed neither to be exact nor to describe the situation properly.) By this example we tried to illustrate that the technology of a product (e.g. electrical machines) relies on a certain background without which the concrete technology - accepted by everyday common sense as practical and useful - is just impossible to create, and this indispensable background consists of conceptual systems and a hierarchy of very abstract and exact theories. In the above example it is worthwile to mention that electrical engineering and theoretical electrical engineering form two independent sciences. Thus the production technology of machines can not be based only on physics, at least two other self-contained sciences are also needed, which in the meantime are not of practical, but of theoretical orientation. For any kind of traditional products it is generally true that within the technical - scientific culture behind their manufacture technology there is a quite rich and complex subculture, which is a network of theories and conceptual systems of different levels of abstraction and orientation. The system of these subcultures came into being continuously in the course of the development of sciences.

The production technology of programs simply misses such a cultural background. Moreover, there is no related area from which to borrow it. To put it more sharply: the programs are an industrial product missing any precursor. The hitherto existing industrial products were always tangible (concrete) physical objects. In contrast, physically a program is nothing but a string of symbols. Therefore those laws which influence the production of programs are of completely different nature than those influencing the manufacture of traditional products. We will mention here only one of such differences. Since programs are only strings of symbols, their production does not require any raw materials or constituents and so no material difficulties can hinder or slow down the production of arbitrarily complex programs. As a consequence, in no time the programmer finds himself involved in an intricate program which he is no more able to keep track of [12]. Today's computer engineering is full of program systems whose complexity is unbelievably greater than that of the artificial systems elaborated ever by our civilization before computer engineering existed. Let us recall those program systems of which we wrote above that they could not be accomplished although, in principle, they were realizable. The real reason why these systems could not be realized was their immense complexity. Thus the only obstacle to realizing the above described programs is pure complexity as such. (A terminological remark: the complexity we are talking about herein has nothing to do with the recent branch of combinatorics called "complexity theory" or sometimes "computational complexity". Our version of complexity was introduced in general system theory and in cybernetics, cf. [2]. We are sorry for this coincidence of names, but we emphasize that there is not even the remotest connection, actually, these are two diametrically opposite research directions.) In the world of programs we have produced (more precisely we were forced to produce) so intricate, richly and many-foldly organized - in one word complex - systems, that, as a consequence, we are facing now a new, vital and self-contained field of problems: that of complex systems. Badly needed is a <u>theory of complex systems</u> as an independent but exact mathematical theory aimed at the production, handling and study of highly complex systems. This theory would disregard the system's origin, its nature etc., the only aspect it would concentrate on would be its being complex.

Let us return to the definition of complexity. A thing is complex if it is composed of a huge number of components of different nature and if these components are related to each other in many and intricate ways. Thus the complexity of a thing grows with the number of its components and with the number and variety of kinds of their relation to each other. As synonims of the word "complex" we can view the words "colourful", "multifold", "rich", or "having many aspects". It is important to emphasize that complexity is different from being complicated. Cf. the above remark about the difference from combinatorical complexity theory. For a more detailed exposition of our view see [2]§1/7. Typically complex systems are e.g. living organisms and societies. One of the characteristics of complex systems is that either they have no subsystems which could be analysed in isolation from the whole or even if they have such, the whole system can not be understood by understanding all of its isolated parts.

Before the appearence of computer engineering no branch of engineering dealt with the production of so complex artificial systems. Moreover, up to now it has been characteristic of our technical - scientific culture that not even the idea of artificial systems whose complexities approach that of living systems (in the above described sense) did arise. This of course set the tone of our scientific culture, its conceptual apparatus, methodology etc. However unbelievable it may sound, in the world of program systems artificial systems which are complex to such a degree did already appear. Moreover, the most burning practical problem of computer technology is to develop a methodology for the production of program systems being complex in the above sense. But the cultural background needed for even approaching this problem is completely missing.

The theory of complex systems would be aimed at approaching "the complex phenomenon" as a whole. Holism is a fundamental principle in the methodology of science. The need for holism is generally recognized as a characteristic feature of the new epoch following the appearance of computers and cybernetics. Holism was not needed before and hence (due to the logic of evolution) it is completely missing from the fabric of hard sciences. This is widely recognized to be the main source of our troubles. In the changed new situation we are forced to elaborate a holist version of hard sciences. Holism is roughly the negation of reductionism. A typical reductionist argument is the following: "The observable universe contains only finitely many elementary particles. Hence it is only finite mathematics which is relevant to understanding our environment including ourselves." Another example "Understanding a living being is nothing but describing precisely the behaviour of elementary particles it consists of". For the holism - reductionism problem as resulting from the advent of computers see [37]pp.9-10(item 1.3.7) where also the need for a new "qualitative" mathematics is indicated e.g. § 1.3.4,8 and the complex programs we mentioned earlier are touched upon e.g. in § 1.3.1-4. For explanation of holism see [20, 3],[38]e.g. p.17. In management science people became aware of these problems a little before the impact of computers [38]. This was brought to our attention by G. Grätzer. On p.36 of the book General System Theory it reads "General system theory ... is a general science of 'wholeness' ... in elaborate form it would be a logico-mathematical discipline". The main aim of this theory of complex systems is to keep intact the system's being rich beyond our being able (or wanting) to describe it. It should develop a cult of not harming the inexhaustable richness of a complex system when talking about it. I.e. when studying a complex phenomenon, we always have to keep in mind that we study this thing from only one of its countless aspects. To this end we have to represent explicitly the aspects we ignored in this specific study. How is it possible <u>technically</u> to represent something we did not take into consideration? The presently common solution to this question is the explicit use of the class of all possibilities. This means that if we say something then we consider and even represent formally <u>all</u> possibilities (structures) compatible with what we said, and we take so seriously not having mentioned other things that we include also those possibilities about which we know (by common sense, by things not mentioned) that they are impossible (but this impossibility does not follow from what we said). Summing up, the "intactness of the whole" can be retained within exact frames by the explicit use of the class of all possibilities. About the success of this approach see [2]§7, [3].

Within mathematics the above principle can be related to the so called axiomatic method. But as shown by the following quotation from Fried[14], it can make a big difference how we use this axiomatic method: " ... mathematics, whose methods are used, among others, in linguistics, biology, economy etc. These areas of applications, however, require a completely "new" mathematics ... In mathematics the axiomatic method is very old. It occurred first in geometry, but this axiomatic method essentially differs from the above formulated requirements. The aim of the axiomatization of geometry was to distinguish a single object (the plane or the space) from anything else. Whereas the essence of the questions raised before is exactly our being able to analyse at the same time more than one - perhaps completely different - cases. Within mathematics the first investigations of this type were performed in the field of abstract algebra. We could even say that it is exactly the starting of such investigations which led to the birth of abstract algebra".

Thus it is <u>not</u> the so called <u>categorical</u> axiom systems which are relevant to the new science, but those of the opposite type (a system of axioms is categorical if there are only a few "possible worlds" in which the axioms hold, i.e. if it describes a fixed reality exhaustively). Euclid, Bolyai, abstract algebra and universal algebra are to illustrate, in this given order, the degree of negating the idea of categorical axiom systems. While Euclid aspired to describe by his axioms a single world as exactly as possible (categoricity!), Bolyai's axioms are already compatible with many possible worlds. Abstract algebra already contains many decidedly not categorical axiom systems, e.g. group theory, lattice theory etc. Universal algebra no more puts up with the study of finitely many fixed axiom systems. Instead, the axiom systems themselves form here the subject of study. In other words, the theorems of universal algebra can be applied to arbitrary axiom systems. Universal algebra investigates those regularities which hold when we try to describe an arbitrary phenomenon by some axioms we ourselves have chosen.

The above give already a hint why the theory of programming needs exactly universal algebra: On one hand, because we cannot expect the structures arising in connection with programming to satisfy the axioms of one of the finitely many fixed axiom systems of abstract algebra. On the other hand - and this is the really important reason - it is the process of abstraction in itself which is of importance in computer science, the rise of axiom systems there being an everyday event whose handling needs the so called metamethods. One of the most outstanding researchers of computer science, J. Goguen writes [16] that it is the mathematical theory and methodology of abstraction which we need the most and we should investigate the exact laws of the process of abstraction, since in the new and changed situation the specialists working in practice (i.e. not only the theoreticians) are supposed to deal with abstraction to an immense extent never heard of before. He illustrates his observation that abstraction has become an obligatory tool even in the routine programming by giving among others the examples of abstract syntax and abstract data types. In the same paper he writes: "Please note that our informal discussion has been motivated by practical issues; it is not 'abstract nonsense', but is aimed at reducing the high cost of software".

We conclude by taking a look at today's mathematics trying to estimate from which of its branches could a new mathematics "of quality" emerge forming the basis of the (not yet existing) culture of complex systems. (More or less this amounts to looking for the mathematics of general system theory (in the sense of Bertalanffy, Ashby and their followers). It is again Goguen who seems to consistently consider general system theory as a deeper background for theoretical computer science.) Taking into account Fried's above quoted remark concerning categoricity, it is universal algebra, category theory, algebraic logic and model theory which seem promising for our problems.

For a more concrete elaboration of the present ideas and for a review of the actual research work done in the above outlined direction the reader is referred to our survey series [1]. Therein we also classified the existing literature into 11 subdivisions and tried to outline the interconnections between them, too. In the list of references we list a few randomly chosen new items.

It is important to note that in the present work (as well as in [1]) universal algebra (UA from now on) is understood in the broader sense. E.g. we include a great portion of model theory. In accordance with the chain of ideas unfolded above, it is the <u>non-categorical</u> part of model theory what we include into UA of the present paper. We include almost every part of model theory that is <u>not</u> oriented towards categoricity. We include <u>partial algebras</u> (and also many-sorted algebras and other generalizations) not only into UA but also into the most essential core of UA, see [7,8]. As a contrast, fashionable recent directions (with really hard results etc.) of UA are not necessarily important from our point of view. A few words in favour of partial algebras: Computer science (CS from now on) needs flexibility. Partial algebras increase flexibility of UA. Further, CS needs insight into the process of abstraction and preferably so that the particular language underlying this process should not be fixed. That is, it would be greatly helpful for CS if UA could be abstracted from its particular language which happens to be equational logic of total algebras. In partial algebra theory you change the logic (or language) of UA changing both the syntax and the semantics (the models are different) of this language. The same is repeated in many-sorted (heterogeneous) algebras and other related fields. (But partial algebras are the most typical example where all the problems arise.) So the partial algebraist faces a dilemma: either we have to repeat the whole of UA in a growing number of different versions (for partial algebras, for many-sorted, etc) or we try to develop a single something that is applicable to all these versions of UA. Even if one is not taking the second choice (explicitly, that is) research in partial algebra theory helps to abstract UA from its traditional language (or logic). Research in the semantical part of CS arrived at a similar conclusion (from a different direction). Namely, Makowsky and Mahr [27,28] found that a kind of "ultra soft model theory" is needed in CS especially in specifications (data, programs etc). This is a UA oriented version of soft model theory (i.e. abstract model theory) which concentrates on logics like equational logic, quasi-equational logic etc. instead of $L_{\kappa\lambda}$ and other strengthenings of $L_{\omega\omega}$. (Note again the shift from categoricity to noncategoricity, i.e. from small axiomatizable classes to large axiomatizable classes; here the shift was brought about explicitly by CS.) These works also provide a kind of feedback from CS to UA.

A few of the recent developments not reported in [1]: There is a recent upheaval of abstract model theory in CS e.g. Goguen-Burstall[17] is an excellent exposition, cf. also [27,28]; well we could not list all the references here for lack of space. We warmly recommend as introductory materials [17,28,16] not only for the topics in their titles but also for the subject matter of the present paper in general. They explain many things we only hinted here. Of these [17] is completely and thoroughly universal algebraic even in the narrower sense. Essential applications of reduced products to CS are e.g. in [10,35,36,32]. Ultraproducts play a major rôle e.g. in comparing the powers of famous program verification methods and in characterizing their implicit information contents. These are two important classical problems of CS the latter being raised e.g. in Burstall[9] setting the course for much of subsequent development in theory of programming. Algebraic logic in CS is gaining even more momentum. (E.g. cylindric-relativized set algebras (called Crs's in [25]), cf. [25,21,41],[1]Part II etc; dynamic algebras of programs the latter being relevant to artificial intelligence too; category theoretic parts of algebraic logic e.g. [17] and related works e.g. in stepwise refinements of specifications etc.) Deep model theoretical tools were used in the computer science papers [19,31] etc.

REFERENCES

[1] Andréka,H. Németi,I.: Applications of universal algebra, model theory and categories in computer science. (Survey and bibliography.) Parts I-II in Comput.Linguist.Comput.Lang. 13(1979)pp. 152-282, 14(1980)pp.43-65. Part III in Fundamentals of Computation Theory'81, LNCS 117, Springer, Berlin, 1981. pp.281-290.

[2] Ashby,R.: An introduction to cybernetics. Chapman & Hall Ltd., 1956.

[3] Ashby,R.: General systems theory as a new discipline. In: General Systems Yearbook of Soc. General Systems Res. Vol 3, 1957, Ann Arbor, Michigan. pp.1-6.

[4] Ashby,R.: Design for an intelligence amplifier. In: Automata Studies (ed. Shannon-McCarthy), Friaction, Princt. Univ. Press, 1956.

[5] Ashby,R.: The set theory of mechanism and homeostasis. Electrical Engineering Research Laboratory, Univ. of Illinois, 1961.

[6] Benecke,K. Reichel,H.: Equational Partiality. Algebra Universalis, to appear.

[7] Burmeister,P.: A unifying approach towards a two-valued (many-sorted) model theory for partial algebras. Algebra Universalis, to appear.

[8] Burmeister,P. Reichel,H.: On the current state and trends in the theory of partial algebras especially for computer scientists. Computational Linguistics Logics Languages, to appear.

[9] Burstall,R.M.: Program proving as hand simulation with a little induction. In: Information Processing'74, North-Holland, 1974. pp.308-312.

[10] Csirmaz,L.: On the strength of "sometimes" and "always" in program verification. Information and Control, to appear.

[11] van Emde Boas,P. Janssen,T.M.V.: Some observations of compositional semantics. In: Logics of programs, LNCS 131, Springer, Berlin, 1982. pp.137-149.

[12] Dijkstra,E.W.: On the interplay between mathematics and programming. In: Program construction, LNCS 69, Springer, Berlin, 1979. pp.35-46.

[13] Francez,N. Rodeh,M.: A distributed abstract data type implemented by a probabilistic communication scheme. Preprint, IBM Israel Scientific Center, Haifa, 1980.

[14] Fried,E.: Abstract algebra in an elementary way. (In Hungarian.) Müszaki Könyvkiadó, Budapest, 1972.

[15] Ganzinger,H.: Programs as transformations of algebraic theories. Inst. für Informatik, Technische Univ. München, 1982.

[16] Goguen,J.A.: Some ideas in algebraic semantics. In: Proc. 3rd IBM Symp. on Mathematical Foundations of Computer Science, Kobe Japan, ~1979.

[17] Goguen,J.A. Burstall,R.M.: Institutions: Abstract model theory for program specifications. SRI International and Univ. of Edinburgh, 1983. pp.1-44.

[18] Guessarian,I.: Applications of universal algebra to computer science. Studia Sci. Math. Hung., to appear.

[19] Hájek,P.: A simple dynamic predicate logic. Theoretical Computer Science, to appear.

[20] Hofstadter,D.R.: Gödel, Escher, Bach: An eternal golden braid. Basic Books Inc., New York, 1979.

[21] Imieliński,T. Lipski,W.Jr.: The relational model of data and cylindric algebras. ICS PAS Report, Warsaw, June 1981.

[22] Janssen,T.M.V.: Foundations and applications of Montague grammar. Mathematisch Centrum, Amsterdam, 1983. vii+440 p.

[23] Kanda,A.: Data types as initial algebras: A unification of Scottery and ADJery. In: 19th Ann. Symp. on Foundations of Computer Science (Oct. 1978, Ann Arbor Michigan), 1978. pp.221-230.

[24] Kfoury,A.J.: Loop elimination and loop reduction - a model theoretic analysis of programs. In: 21st Ann. Symp. on Foundations of Computer Science (Oct. 1980, Syracuse New York), 1980. pp. 173-184.

[25] Knuth,E. Rónyai,L.: Closed convex reference schemes. (A junction between computer science, cylindric and partial algebras.) Proc. IFIP Conf. (1983 Hungary).

[26] Lipton,R.J.: Model theoretic aspects of computational complexity. In: 19th Ann. Symp. on Foundations of Computer Science (Oct. 1978, Ann Arbor Michigan), 1978. pp.193-200.

[27] Mahr,B. Makowsky,J.A.: An axiomatic approach to semantics of specification languages. Computational Linguistics Logics Languages, to appear.

[28] Makowsky,J.A.: Model theoretic issues in theoretical computer science, Part I: Relational data bases and abstract data types. Technion Israel Inst. of Technology, Dept. of Computer Science, TR No 270, 1983.

[29] Makowsky,J.A.: Measuring the expressive power of dynamic logic: an application of abstract model theory. In: Proc. of 7th ICALP, LNCS 85, Springer, Berlin 1980. pp.409-421.

[30] Meseguer,J.: A Birkhoff-like theorem for algebraic classes of

interpretations of program schemes. In: Formalization of programming concepts, LNCS 107, Springer, Berlin, 1981.

[31] Meyer,A.R. Parikh,R.: Definability in dynamic logic. In: Proc. 12th ACM Symp. on Theory of Computing (1979). pp.167-175.

[32] Németi,I.: Nonstandard dynamic logic. In: Logics of programs, LNCS 131, Springer, Berlin, 1982. pp.311-348.

[33] Parikh,R.: Propositional dynamic logics of programs: A survey. In: Logic of programs, LNCS 125, Springer, Berlin, 1981. pp. 102-141.

[34] Parikh,R.: Some applications of topology to program semantics. In: Logics of programs, LNCS 131, Springer, Berlin, 1982. pp. 375-386.

[35] Sain,I.: Successor axioms increase the program verifying power of full induction. Math. Inst. Hung. Acad. Sci., Budapest 1983.

[36] Sain,I.: The implicit information content of Burstall's (modal) program verification method. Math. Inst. Hung. Acad. Sci., Budapest, 1983.

[37] Sloman,A.: The computer revolution in philosophy. The Harvester Press, Sussex, 1978.

[38] Systems thinking. Penguin modern management readings, 1969.

[39] Thatcher,J.W. Wagner,E.G. Wright,J.B.: Notes on algebraic fundamentals for theoretical computer science. Mathematical Centre Tract 109, Amsterdam 1979. pp.83-163.

[40] Wirsing,M. Broy,M.: An analysis of semantic models for algebraic specifications. In: Proc. International Summer School on Theoretical Foundations of Programming Methodology, Münich 1981. 66 p.

[41] Zlatos,P.: On conceptual completeness of syntactic - semantical systems. Computational Linguistics Logics Languages, to appear.

Mathematical Institute of the
Hungarian Academy of Sciences
Budapest 1376, Pf. 428, Hungary.

Universal Algebra and its Links with Logic,
Algebra, Combinatorics, and Computer Science
Proc. "25. Arbeitstagung über Allgemeine Algebra", Darmstadt 1983
P. Burmeister et al. (eds.)

216 – 229

THE COMPUTATIONAL COMPLEXITY OF SOME PROBLEMS OF UNIVERSAL ALGEBRA

L. Kučera, V. Trnková

I. Introduction. In this paper, we deal with some problems, encountered in universal algebra, from the point of view of their computational complexity. The question of the existence of an algorithmic solution of problems has a long tradition in universal algebra, let us mention e.g. an effort devoted to the word problem in varieties of algebras.

We do not investigate algorithmic solvability of algebraic problems; all the questions investigated here can be solved in an algorithmic way evidently (let us say explicitly that the word "algebra" always means a finite algebra of a finite finitary type in this paper) and it is even seen at the first sight how an algorithm for the solution could be constructed. However, algorithms of this type, usually called "brute force algorithms" often run in the time growing exponentially with the size of algebras in question and therefore they are useless from the practical point of view even for relatively small objects. There is an obvious problem to find a faster algorithm (or to prove its non-existence).

In some cases, the brute force algorithm is sufficiently fast, e.g. if we test the associativity of a given binary operation, then it is sufficient to take into account all the triples of elements of the algebra in question. But also in this case, it is reasonable to ask whether the brute force gives an optimal method, e.g. whether it is really necessary to inspect all the triples.

As a formal description of models of computation we can use either Turing machines or random excess machines (RAM), see [AHU]. The former is simpler, the latter is closer to existing computers.

We say that a problem can be solved in polynomial time if there is a polynomial p and an algorithm $\mathcal{A}$ (e.g. Turing machine or

RAM) such that $\mathcal{A}$ solves the problem for all possible input data and the computation always stops after at most p(n) steps, where n is the size of the input data, i.e. of a description of the objects in question. (Though an algebra of a fixed type Δ on an n-point set is given by the table of the results for each operation $\sigma \in \Delta$, i.e. by $d = \sum_{\sigma \in \Delta} n^{|\sigma|}$ data, its size is usually taken as the number of its elements; this does not influence the above definition because d is a polynomial in n.) The class of all problems solvable in the polynomial time is denoted by $\mathcal{P}$ and is usually considered to be the class of really tractable problems. Unfortunately, there are many natural problems not believed to belong to the class $\mathcal{P}$ (though this fact is often neither proved nor disproved). There is another important class of problems, the class $\mathcal{NP}$ which contains almost all problems usually encountered. $\mathcal{NP}$ is the class of problems solvable nondeterministically in polynomial time. The precise definition of the class $\mathcal{NP}$ can be found e.g. in [AHU]. Instead of this, let us explain an intuitive meaning of this notion. Consider classes $\mathcal{C}_1$, $\mathcal{C}_2$ of mathematical objects and a predicate π on $\mathcal{C}_1 \times \mathcal{C}_2$; given $\sigma \in \mathcal{C}_1$, we have to decide whether there is $\sigma' \in \mathcal{C}_2$ such that $\pi(\sigma, \sigma')$ holds. If the validity of π can be tested in a polynomial time then we can "solve" our problem by guessing at an object σ' and then checking $\pi(\sigma, \sigma')$. In other words, the determining whether

$$\exists \sigma' \pi(\sigma, \sigma')$$

is true or not, can be very difficult; but if the solution is affirmative, we can prove its correctness in the polynomial time by simply giving the object σ' and verifying $\pi(\sigma, \sigma')$. E.g. we can prove that two given algebras of a given type Δ are isomorphic by describing a particular bijection and proving that this is an isomorphism; or, more in general, we can prove the existence of a homomorphism of an algebra A into an algebra B by describing a particular mapping of A into B and proving that it is a homomorphism (here, $\mathcal{C}_1$ is the class of all pairs (A,B) of algebras of the type Δ , the size is taken e.g. as n = max (card A, card B), $\mathcal{C}_2$ is the class of all mappings). Note that the definition is asymmetric; we don't know any simple proof of the fact that $\exists \sigma' \pi(\sigma, \sigma')$ is not true if this is the case. Problems like the testing of an existence of a subset, mapping, bijection etc. with some simple property belong to $\mathcal{NP}$. It is obvious that $\mathcal{P} \subseteq \mathcal{NP}$; the question whether $\mathcal{P} \neq \mathcal{NP}$ is open and it is one of the most famous and important questions in contemporary mathematics. It is commonly believed that the inequation holds (or is at least consistent with the axioms of the Peano arithmetic, see e.g. the survey paper [JY]).

The next important notion is a polynomial reduction. For the precise definition, see again [AHU] or [GJ], we describe its intuitive meaning. Consider classes $\mathcal{C}_1$, $\mathcal{C}_2$ and properties P_1, P_2. A problem Π_i, i = 1,2, is to decide whether a given object $\sigma \in \mathcal{C}_i$ has the property P_i. We say that Π_1 is polynomially reducible to Π_2 and write $\Pi_1 \propto \Pi_2$ if there is a mapping $f: \mathcal{C}_1 \rightarrow \mathcal{C}_2$ such that

a) for every $\sigma \in \mathcal{C}_1$, σ has the property P_1 iff $f(\sigma)$ has the property P_2 and
b) f can be computed by an algorithm running in polynomial time (i.e. there is a polynomial p such that $f(\sigma)$ is computed in at most p(n) steps, where n is the size of a description of the object σ).

If $\Pi_1 \propto \Pi_2$ then we can test whether $\sigma \in \mathcal{C}_1$ has the property P_1 by constructing (in a simple way) $f(\sigma)$ and testing whether $f(\sigma)$ has the property P_2. We say that Π_1 and Π_2 are polynomially equivalent iff $\Pi_1 \propto \Pi_2$ and $\Pi_2 \propto \Pi_1$.

In [C], S. Cook proved the following very important theorem: There exist problems $\Pi \in \mathcal{NP}$ such that for each $\Pi' \in \mathcal{NP}$, $\Pi' \propto \Pi$.

Such problems are called $\mathcal{NP}$-complete and represent the largest possible computational complexity in the class $\mathcal{NP}$ (up to the polynomial equivalence). Till now, several hundreds of $\mathcal{NP}$-complete problems have been found (see [GJ]), among them the problem of 3-colorability of a given graph, i.e. the question whether there is a compatible mapping of a graph G into the complete graph on 3 vertices. If Δ is a given type of universal algebras with $\sum \Delta \geq 2$ (i.e. Δ contains at least two unary operations, or at least one at least binary operation), then there is an easy constructible full embedding of the category of graphs and their compatible mappings into the category Alg (Δ) of all universal algebras of the type Δ (see e.g. [HP]). This gives a polynomial reduction of the problem of the existence of compatible mappings of graphs (the special case of which is the 3-colorability) into the problem of the existence of a homomorphism between two algebras of the type Δ ; hence, given algebras A, B of the type Δ , the testing of the existence of a homomorphism of A into B is $\mathcal{NP}$-complete.

In the part II. of the present paper, we examine the complexity of the isomorphism testing in varieties of universal algebras. The last part III. is devoted to the testing of the first order properties. Here, the brute force algorithm (sketched in III.) is always polynomial. However, there are remarkable algorithms for some special cases, particularly for the testing whether a given groupoid is a group or whether a given bigroupoid (= an algebra with two binary operations) is a lattice; in both these cases, the associativity of the binary operations in question can be tested faster than the inspection of all the triples of elements of the tested algebra.

II. Testing of isomorphism of algebras. The testing whether two given mathematical structures are isomorphic or not has received much attention. Originally, the problem was to decide about the existence of an isomorphism of graphs (we always mean finite graphs). The interest in the problem stems in part from its unresolved complexity: this problem is in the class $\mathcal{NP}$ but it is not known either whether it is $\mathcal{NP}$-complete, or belongs to the class $\mathcal{P}$ or even easier than $\mathcal{NP}$-complete problems but still not in $\mathcal{P}$.

If G, G′ are two graphs on an n-point set V then the brute force algorithm decides whether they are isomorphic or not as follows: the testing whether a given bijection of V onto itself is an isomorphism of G onto G′ takes $O(n^2)$ steps (let us recall that given functions f, g, we write $f = O(g)$ if there are constants K and n_0 such that for any $n \geq n_0$, $f(n) \leq K \cdot g(n)$), and in the worst case the algorithm has to inspect all these bijections - there are n! of them. Hence this algorithm solves the problem in $O(n^2 \cdot n!)$ steps. The fastest known algorithm for this problem has been created by V.N. Zemlyachenko and L. Babai, [B1]. This deep and witty algorithm tests the existence of an isomorphism of two graphs with n vertices in $O(2^{n^{2/3+o(1)}})$ steps (where $o(1)$ denotes a function of n with $\lim_{n \to \infty} o(1) = 0$). Let us notice that for some classes of graphs, the isomorphism testing has been proved to be in the class $\mathcal{P}$; we mention here at least the class of all graphs with valences bounded by a given constant (see [L]). On the other hand, in a number of classes of graphs (like bipartite graphs, regular graphs, Hamiltonian graphs and many others), the isomorphism testing is still as complicated as in the class of all graphs. Problems polynomially equivalent with the isomorphism testing of graphs are called <u>isomorphism complete.</u> A survey of results about isomorphism complete problems is given in [BC].

Let us discuss here the <u>isomorphism testing of universal algebras.</u> For any type Δ with $\Sigma\Delta \geq 2$, the isomorphism testing in the class of all algebras of the type Δ is isomorphism complete. This is an immediate consequence of a more general construction in [HP]. The result is stated explicitly in [M 1]. If $\Sigma\Delta \leq 1$, then the isomorphism testing in the class of all algebras of the type Δ is in $\mathcal{P}$. This follows by an easy modification of the algorithm for the isomorphism testing of trees described e.g. in [AHU].

Since the isomorphism testing in the whole class of all algebras of the type Δ is isomorphism complete (whenever $\Sigma\Delta \geq 2$), there is an interesting field of problems which varieties are already small enough such that the isomorphism testing in them is subexponential or even in $\mathcal{P}$. Let us mention that, for example, the variety $\mathbb{V}(\varphi,\psi)$ of all unary algebras with two idempotent commuting operations φ, ψ (given by the equation $\varphi\psi(x) = \psi\varphi(y)$ and $\varphi^2(x) = \varphi(x)$, $\psi^2(x) = \psi(x)$ is still large - the isomorphism testing in it isomorphism complete. In [KT 1], a polynomial reduction of the isomorphism testing of graphs into the isomorphism testing in $\mathbb{V}(\varphi,\psi)$ is presented, we visualize it below for G being the chain on three vertices.

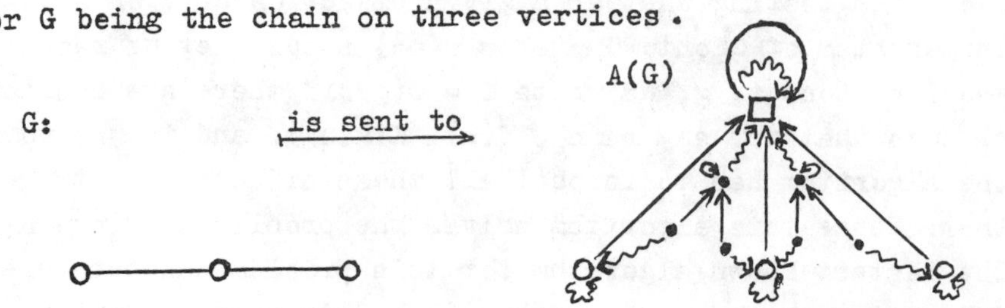

(In the picture of A(G), the arrows ⟶ and ⇝ indicate the operations φ and ψ ; A(G) is created such that we add one fix element to the set of vertices of G and replace each edge of G by the configuration which can be seen from the picture.) Clearly, two graphs G, G′ are isomorphic iff A(G) and A(G′) are isomorphic.

A complete discussion of all varieties of unary algebras with arbitrary (finite) number of operations is presented in [KT 2], where an algorithm $\mathcal{A}$ is constructed such that given two unary algebras A, B, then either

(i) $\mathcal{A}$ tests whether they are isomorphic or not in a polynomial time or

(ii) the isomorphism testing in any variety containing A or B is isomorphism complete.

(The corresponding polynomial reductions are the embeddings of categories constructed in [S].) Hence the isomorphism testing in any variety of unary algebras is either isomorphism complete or is in $\mathcal{P}$. The description which of these two cases takes place for a given variety $\mathcal{V}$ is also described in [KT 2].

If the type Δ of algebras consists of unary and nullary operations, it can be reduced to the previous case because each nullary operation can be replaced by a unary operation and the equation

$$\sigma(x) = \sigma(y).$$

Let us move to binary operations. First, we mention that the isomorphism testing of Abelian groups is in $\mathcal{P}$: the classical theorem (see e.g. [F]) says that each finite Abelian group G can be expressed as $G = C_{k_1} \oplus \dots \oplus C_{k_s}$, where C_{k_i} are cyclic groups and k_i divides k_{i-1}. The proof of this theorem is constructive, it gives directly the method how to construct this direct decomposition and it can be seen that it can be done in a polynomial time. And two groups $G = C_{k_1} \oplus \dots \oplus C_{k_s}$ and $G' = C_{k'_1} \oplus \dots \oplus C_{k'_{s'}}$ are isomorphic iff $(k_1, \dots, k_s) = (k'_1, \dots, k'_{s'})$.

It is not known, whether the isomorphism of two general groups can be tested in polynomial time; nevertheless, we can do it in subexponential time, i.e. faster than any known isomorphism testing of graphs. In [T], an algorithm is described which for each group G with n elements constructs a set of generators Γ such that card $\Gamma \leq \log n$ (log n always means the dyadic logarithm) and each element is already given as an expression in elements of Γ (we describe such an algorithm with some further properties in III.) - this is all done in $O(n^2)$ steps. This algorithm is used for the isomorphism testing of groups G, G′: first, the above set of generators Γ is constructed; the test whether a given map m: $\Gamma \longrightarrow G'$ can be extended to an isomorphism of G onto G′ can be done in $O(n^2)$ steps and there are $n^{\log n}$ such maps m; hence the isomorphism of groups can be tested in a subexponential time $O(n^{c_1 \log n + c_2})$, where c_1, c_2 are constants (for some problems of the same complexity - see [M 2]).

On the other hand, the isomorphism testing in the class of all commutative semigroups is known to be isomorphism complete, see [Bo] . The polynomial reduction of the isomorphism testing of graphs into the isomorphism testing of commutative semigroups (in fact of upper semilattices) is visualized in the following picture (where analogous conventions as in the previous picture are used).

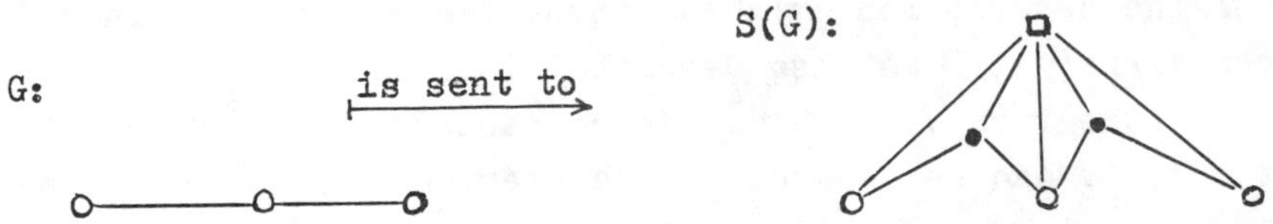

A complete discussion of the isomorphism testing in all pseudovarieties of semigroups and monoids is given in [GGK 2] and [GGK3]. (Pseudovarieties are called varieties by S. Eilenberg in [E]; these are classes of algebras closed with respect to subalgebras, homomorphic images and finite products.) In [GGK 2], a pseudovariety $\mathcal{V}$ is called critical if the isomorphism testing in any variety properly contained in $\mathcal{V}$ can be computed in a polynomial time but in $\mathcal{V}$ this is isomorphism complete. In [GGK 2], all critical pseudovarieties of semigroups are described and the following deep result is proved in the full version [GGK 3]: every pseudovariety of semigroups either contains one of the critical pseudovarieties (and then the isomorphism testing in it is isomorphism complete) or the isomorphism testing in it can be computed in $O(n^{c_1 \log n + c_2})$ steps. The situation in the monoid pseudovarieties is simpler: every pseudovariety of monoids which is not contained in the class of all groups contains the pseudovariety of all commutative idempotent monoids - the only critical pseudovariety of monoids. Much attention has been paid to the isomorphism testing of distributive lattices . It is still open whether this problem is in $\mathcal{P}$; anyway, an algorithm for the isomorphism testing of distributive lattices is known, faster than the known algorithms for the isomorphism testing of graphs. In [B 1] L. Babai proved the following result: if G, G′ are graphs on an n-point set with a coloring of vertices (this is here a mere decomposition of the set of vertices into classes "colored by the same color") and if the valence of any vertex v in each color is at most k (i.e. for each i the number of all neighbours of v colored by the i-th color is at most k) then the existence of a color-preserving isomorphism can be tested in less than $n^{k^2(c\log k + \log n)}$ steps, where c is a constant. If L and L′ are distributive lattices with n elements, let us denote by P and P′ the posets of all their join-irreducible elements. The covering relation decomposes P and P′ into levels of incomparable elements with at most log n elements (x is in the i-th level iff it covers an element of the (i-1)-th level and there is no element between x and any element of the (i-1)-th level). Since any isomorphism of L onto L′ has to preserve these levels, we color the i-th level of P and P′ by the i-th color; hence the above estimate can be used with k = log n, so that the isomorphism of P and P′ (and hence of L and L′) can be tested in $O(n^{d \log^3 n})$, where d is a constant (a probabilistic algorithm which tests it in $O(n^{c \log \log n})$ with the probability $> \frac{1}{2}$ is constructed in [B 2]).

On the other hand, the isomorphism testing in the class of <u>all lattices</u> is isomorphism complete. The polynomial reduction of the isomorphism testing for graphs into isomorphism testing of lattices is obtained by filling up the upper semilattices in the picture above by the zero-element (see [Bo]).

In this field of problems, there are many open questions. Can the isomorphism of graphs be tested faster? (This would influence all the isomorphism complete problems, the negative answer would imply $\mathcal{P} \neq \mathcal{NP}$.) Can the isomorphism of groups or distributive lattices be tested in a polynomial time? (The negative answer would imply $\mathcal{P} \neq \mathcal{NP}$ again.) Nothing is known about modular lattices: is the isomorphism testing in this class isomorphism complete? or can it be done in subexponential time? or even in polynomial time? The isomorphism testing in varieties of algebras of higher arities has not yet been attacked.

<u>III. The first order properties.</u> In this part, we shall deal with the testing of properties of algebras given by formulas of the first order language. As we sketch below, the brute force algorithm

for such property always tests its validity in a given algebra in polynomial time. We are again interested in the problem to test it as fast as possible. Since we deal with problems in $\mathcal{P}$, we use RAM as a model of computation. (Any n-step computation of RAM can be simulated by a Turing machine in $O(n^2)$ steps, see e.g. [AHU]; Turing machines are used to obtain rough estimation of the time complexity - as in the part II. where the time complexity was investigated only up to polynomial equivalence - while RAM is used to evaluate the complexity more precisely. RAM admits to reach any datum in its memory in one step; for more about the data structure of RAM and its computational techniques see [AHU].)

Let us sketch briefly how the brute force algorithm works. Let Δ be a type of universal algebras and $L(\Delta)$ denote the corresponding first order language (one can admit also other predicates than $=$, if necessary). Let Φ be an arbitrary formula of $L(\Delta)$, $(x_{i_1},\dots,x_{i_k})$ the list of all its free variables. Given an algebra A of the type Δ , let us denote by Φ_A the set of all k-tuples $(a_1,\dots,a_k)$ of elements of A such that $\Phi(a_1,\dots,a_k)$ is valid. By the definition of formulas (see e.g. [G]), Φ is either an atomic formula or Φ is $\neg\Psi$ or $\exists x_j \Psi$ or $\Phi^1 \vee \Phi^2$. This gives directly a way, how to compute Φ_A . Let us suppose that card A = n and Φ has precisely k free variables $x_{i_1},\dots,x_{i_k}$. If Φ is an atomic formula, then Φ_A can be computed in $O(n^k)$ steps, evidently. If Φ is $\neg\Psi$ and Ψ_A can be computed in $O(f(n))$ steps, then Φ_A can be computed in $O(f(n)+n^k)$ steps. Let Φ be $\exists x_j \Psi$; if x_j has no free occurence in Ψ , then $\Phi_A = \Psi_A$, else $\Psi_A \subseteq A^{k+1}$ and Φ_A can be computed from Ψ_A by the missing of the coordinate corresponding to x_j, i.e. in $O(n^{k+1})$ steps. Finally, if Φ is $\Phi^1 \vee \Phi^2$ and Φ_A^i can be computed in $O(f_i(n))$ steps, then Φ_A can be computed in $O(f_1(n)+f_2(n)+n^k)$ steps. We conclude that for each formula Φ,

Φ_A can be computed in $O(n^s)$ steps where s is the number of all variables (bounded or free) which occurs in Φ.

However, this estimation is rather rough. Sometimes, even the brute force algorithm can be faster, e.g. the replacing of $\exists x_j \Psi$ by Ψ if x_j has no free occurence in Ψ or renaming a bounded variable in an admissible way can give an equivalent formula with a smaller number of variables, but this change does not influence the algorithm. On the other hand, for the formula

$$\forall x \forall y \forall z\ ((x \otimes y) \otimes z = x \otimes (y \otimes z)),$$

describing the associativity of a binary operation $\otimes$, this estimation is exact. And let us state that for this formula, the brute force algorithm is the fastest one known. On the other hand, faster algorithms have been created, which test the associativity simultaneously with some additional properties. Let us sketch here intuitively (and hence more transparently) the nice and witty algorithm of J. Vuillement (see [V]), which recognizes whether a groupoid on an n-point set is a group in $O(n^2)$ steps.

It is easy to see that, in a given groupoid $G = \{g_1, \dots, g_m\}$ (with an operation $\otimes$, given by a table), the existence of the unit e can be tested in O(n) steps (first, we test whether there is precisely one idempotent in G by inspection of the diagonal of the $n \times n$ matrix of the results of $\otimes$: if not, the algorithm stops and gives the result that G is not a group; if so, we inspect the line and the column in the matrix). The left and the right cancellation can be tested in $O(n^2)$ steps: we inspect any line and any column of the matrix whether no element of G appears in it twice. The main part of the algorithm is the testing of the associativity. It is based on a construction (in $O(n^2)$ steps) of a set $\Gamma = \{g_{i_1}, \dots, g_{i_k}\}$ of generators (more precisely, a list of generators) and mappings γ, δ of $\{1, \dots, n\}$ into itself and a permutation π of $\{1, \dots, n\}$ such that

$$g_{i_1} = e,\ \pi(i_1) = 1,\ \gamma(i_1) = i_1 = \delta(i_1);$$

$$\text{if } g_i \in G \setminus \{e\}, \text{ then } g_i = g_{\gamma(i)} \otimes g_{\delta(i)},\ g_{\gamma(i)} \in \Gamma \text{ and } \pi(\delta(i)) < \pi(i).$$

The algorithm creates successively the list $\Gamma = \{g_{i_1}, \dots, g_{i_k}\}$ and the list L of all elements of G as follows: first, it puts $g_{i_1} = e$ on the first place of Γ and define $\gamma(i_1) = i_1 = \delta(i_1)$; then it puts g_{i_1} on the first place of L and defines $\pi(i_1) = 1$ (π will always denote the number when an element of G appears in L); if $\Gamma_{\ell-1} = \{g_{i_1}, \dots, g_{i_{\ell-1}}\}$ have been already put on the first $\ell-1$ places of Γ and the elements of $L_{\ell-1}$ generated by $\Gamma_{\ell-1}$ as described below have been already put on the initial segment of L, we choose arbitrarily $g_{i_\ell} \in G \setminus L_{\ell-1}$ and write it at the end of $\Gamma_{\ell-1}$; we define $\gamma(i_\ell) = i_\ell$, $\delta(i_\ell) = i_1$ (i.e. $\pi(\delta(i_\ell)) = 1$); we create a continuation L' of the list $L_{\ell-1}$ such that we write g_{i_ℓ} at the end of $L_{\ell-1}$ and it plays a rôle of "father": we multiply it from the left and from the right successively by elements of $\Gamma_\ell = \{g_{i_1}, \dots, g_{i_\ell}\}$

and the newly obtained elements of G (i.e. not yet occuring on the list) are written at the end of this continuation L′ (this determines π !); then the element next to g_{i_ℓ} in L′ takes the rôle of "father", then the next one, and so on; if this procedure stops (i.e. the last "father" does not create any new element), we obtain the list L_ℓ of all elements generated by Γ_ℓ . During this procedure, the functions γ and δ are extended as follows: if g_j starts to be the "father", we have already defined $g_{\gamma(j)} \in \Gamma$ and $g_{\delta(j)}$; if we multiply g_j by $g_{i_y} \in \Gamma_\ell$ from the left and a new element $g_z = g_{i_y} \otimes g_j$ is obtained, the definition of $\gamma(z)$ and $\delta(z)$ is straightforward: we put $\gamma(z) = i_y$ ($= \gamma(i_z)$) and $\delta(z) = j$; if we multiply it from the right, i.e. $g_z = g_j \otimes g_{i_y}$, we have to test the associativity

$g_{\gamma(j)} \otimes (g_{\delta(j)} \otimes g_{i_y}) \overset{?}{=} (g_{\gamma(j)} \otimes g_{\delta(j)}) \otimes g_{i_y}$ $(= g_j \otimes g_{i_y} = g_z)$.

If this does not hold, the algorithm stops and gives the result that G is not a group; if it holds, then we put $\gamma(z) = \gamma(j)$ and, since $\pi(\delta(j)) < \pi(j)$, the element $g_{\delta(j)} \otimes g_{i_y}$ is already on the list of the previously generated elements, say $g_{\delta(j)} \otimes g_{i_y} = g_x$; and we put $\delta(z) = x$. If $G \setminus L_\ell$ is still non-empty, we choose $g_{i_{\ell+1}}$ in it and the procedure continues.

Let us mention that if G is a group, then L_ℓ is its subgroup generated by Γ_ℓ and necessarily card $\Gamma \leq 1+\log n$ (indeed, if $\Gamma = \Gamma_k = \{g_{i_1}, \dots, g_{i_k}\}$, then the subgroup L_{k-1} generated by Γ_{k-1} fulfils card $L_{k-1} \leq \frac{n}{2}$ because $a \otimes g_{i_k} \notin L_{k-1}$ for all $a \in L_{k-1}$; then card $L_{k-2} \leq \frac{n}{4}$ and so on); hence the algorithm stops if the number of elements of Γ_ℓ becomes greater than $1+\log n$ -then G is not a group. Notice that if the algorithm does not stop sooner (giving the result that G is not a group) and really creates $\Gamma, \gamma, \delta, \pi$, it does so in $O(n^2)$ steps: every pair (g_i, g_j) of elements of G is processed at most once. The associativity in the cases

$$g_i \otimes g_j = g_{\gamma(i)} \otimes (g_{\delta(i)} \otimes g_j) \text{ for all } g_i, g_j \in G$$

(this takes $O(n^2)$ steps) and

$$(g_i \otimes g_j) \otimes g_k = g_i \otimes (g_j \otimes g_k) \text{ for all } g_j \in G,\ g_i,\ g_k \in \Gamma$$

(this takes $O(n \log^2 n)$ steps)

is sufficient for the associativity of all the triples (g_i, g_j, g_k) with $g_k \in \Gamma$ (the proof by induction in $\pi(i)$ is quite easy) and this is well-known to be sufficient for the associativity of all triples of elements of G. Hence the final result whether G is a group or not is obtained in $O(n^2)$ steps.

Other nice and witty algorithms have been presented in [GGK 1] [GGKR],[T], recognizing e.g. whether a groupoid with n elements is

- an Abelian group - in $O(n^2)$ steps,
- a solvable group - in $O(n^2)$ steps,

or whether a given bigroupoid with n elements is

- a ring - in $O(n^2)$ steps,
- an associative ring - in $O(n^2)$ steps.

Let us describe the results concerning lattices. By [GGK 1], it can be recognized whether a lattice is modular in $O(n^2)$ steps and it can be recognized whether a bigroupoid is a distributive lattice in $O(n^2)$ steps; on the other hand, the fastest known algorithm for the testing whether a bigroupoid is a lattice recognizes it in $O(n^{\frac{5}{2}})$ steps, see [GGKR].

Finally, let us mention some other problems encountered in universal algebra (not necessarily in the first order language) which seem to be interesting,e.g. the testing whether a given algebra has its automorphism group or its congruence lattice isomorphic to a given one. A special case: whether a given algebra is automorphism free or simple. While the first problem is isomorphism complete and its discussion in varieties of universal algebras is closely related to the isomorphism testing, a polynomial algorithm has been created for the testing whether a given algebra is simple: it can be done in $O(n^{k+1} + n^2 \cdot \log n)$ steps, where k is the maximal arity of the type of the tested algebra, see [DDK 1] (and the full version [DDK 2]; in these papers, also the subdirect irreducibility is investigated; it can also be tested in $O(n^{k+1}+n^2 \cdot \log n)$ steps). For more general lattices L, the testing whether the congruence lattice of a given algebra is isomorphic to L has not been solved.

Acknowledgment. We are grateful to our colleagues Ala Goralčíková and Václav Koubek for fruitful discussions. Václav Koubek and Jiří Adámek also read the manuscript and we are indebted to them for some critical comments (but they are not responsible for the inaccuracies in our writing).

References

[AHU] A.V. Aho, J.E. Hopcroft, J.D. Ullman: The design and analysis of computer algorithms, Addison-Wesley, Reading, Mass., 1974.

[B 1] L. Babai: Moderately exponential bound for graph isomorphism, Proceedings of FCT 81, Lect. N. in Comp.Sci.117, Springer Verlag 1981, 34-50.

[B 2] L. Babai: Monte-Carlo algorithms in graph isomorphism testing, preprint.

[B K.S. Booth: Isomorphism testing for graphs, semigroups and finite automata are polynomially equivalent problems, SIAM Journal on Computing 7(1978), 273-279.

[BC] K.S.Booth, C.J.Colbourn: Problems polynomially equivalent to graph isomorphism, Tech.Rep.CS-77-04, U.of Waterloo, June 1979.

[C] S.A. Cook: The complexity of theorem proving procedures, Proc. 3rd Annual ACM Symposium on Theory of Computing, 1971, 151-158.

[DDK 1] M. Demlová, J. Demel, V. Koubek: Several algorithms for finite algebras, Proceedings of FCT 79, Akademie-Verlag Berlin 1979, 99-104.

[DDK 2] M. Demlová, J. Demel, V. Koubek: Algorithms deciding subdirect irreducibility of algebras, to appear.

[E] S. Eilenberg: Automata, languages and machines, Vol.B., Academic Press 1976.

[F] L. Fuchs: Abelian groups, Budapest 1958.

[GJ] M.R. Garey, D.S. Johnson: Computers and Intractability - a guide to the Theory of NP-completeness, W.H. Freeman and Company, San Francisco 1979.

[GGK 1] A. Goralčíková, P. Goralčík, V. Koubek: Testing of properties of finite algebras, Proceedings of ICALP 80, Lect. N. in Comp.Sci. 85, Springer Verlag 1980, 273-281.

[GGK 2] A. Goralčíková, P. Goralčík, V. Koubek: A boundary of isomorphism completeness in the lattice of semigroup pseudovarieties, Proceedings of ICALP 82, Lect. N. in Comp.Sci.140, Springer Verlag 1982, 292-299.

[GGK 3] P. Goralčík, A. Goralčíková, V. Koubek: How much semigroup structure is needed to encode graphs, to appear.

[GGKR] P. Goralčík, A. Goralčíková, V. Koubek, V. Rödl: Fast recognition of rings and lattices, Proceedings of FCT 81, Lect. N. in Comp.Sci. 117, Springer Verlag 1981, 137-145.

[G] G. Gratzer: Universal Algebra (2nd ed.), D. van Nostrand, Springer Verlag 1979.

[HP] Z. Hedrlín, A. Pultr: On full embeddings of categories of algebras, Illinois J. Math. 10(1966), 392-405.

[JY] D. Joseph, P. Young: A survey of some recent results on computational complexity in weak theories of arithmetic, Proceedings of MFCS 81, Lect.N. in Comp.Sci. 118, Springer Verlag 1981, 46-57.

[KT 1] L. Kučera, V. Trnková: Isomorphism completeness for some algebraic structures, Proceedings of FCT 81, Lect.N.in Comp.Sci. 117, Springer Verlag 1981, 218-225.

[KT 2] L. Kučera, V. Trnková: Isomorphism testing of unary algebras, preprint, Charles University, Prague, December 1982.

[L] E.M. Luks: Isomorphism of graphs of bounded valence can be tested in polynomial time, Proc. 21st IEEE FOCS Symp. 1980, 42-49.

[M 1] G.L. Miller: Graph isomorphism: general remarks, Proc. 9th Annual ACM Symposium on Theory of Computing 1977, 143-150.

[M 2] G.L. Miller: On the $n^{\log n}$ isomorphism technique, Proc. 10th Annual ACM Symposium on Theory of Computing, 1978, 51-58.

[RC] R.C. Read, D.G. Corneil: The graph isomorphism disease, J. of Graph Theory 1(1977), 339-363.

[S] J. Sichler: Group-universal unary varieties, Alg.Universalis 11(1980), 12-21.

[T] R.E. Tarjan: Determining whether a groupoid is a group, Inf. Proc. Letters 1(1972), 120-124.

[V] J. Vuillement: Comment verifier si une boucle est un groupe, Technical Report, Université de Paris-Sud, 1976, 1-12.

Luděk Kučera and Věra Trnková

Charles University, Prague, Czechoslovakia

Universal Algebra and its Links with Logic,
Algebra, Combinatorics, and Computer Science
Proc. "25. Arbeitstagung über Allgemeine Algebra", Darmstadt 1983
P. Burmeister et al. (eds.)

230 – 240

PARTIAL ALGEBRAS – A SOUND BASIS FOR STRUCTURAL INDUCTION

H. Reichel

1. Motivations

Structural induction has proved to be a powerful tool in computer science (see [Bur'69],[ADJ'78]). In the development of that so-called 'initial approach' one can notice a similarity to recursion theory. Recursion became a sound concept only in the context of partial functions. We want to demonstrate that to some extent the same is true for structural induction.

It is the aim of this paper to point out formal and informal reasons for the extension of structural induction from total algebras to partial algebras. Some of these reasons are:

(1) The concept of attributed grammars implicitely assumes that structural induction can be performed even if the applicability of production rules is restricted by context conditions.

(2) In algebraic specifications of abstract data types very naturally partial operations appear. It should not be decided by the used specification language wether a naturally partial fundamental operation will be made to a total one by means of so-called error values. It should be a decision of the designer of a software project wether a partial fundamental operation will be completed by an error value or if it will be supplemented by a domain condition. Therefore the semantics of a specification language should provide partial operations. Since the structural behaviour of partial algebras can not be simulated by total algebras, and since the concepts of initial and final algebras strongly depend on the structural behaviour it is not only a matter of tast wether partial operations are available or not.

(3) Readability (clarity) and usability of specifications are to some extent contrary requirements. This causes the necessity of stepwise functional enrichments of specifications. The first

version of a specification should be minimal, i.e. it should only contain the generators. By following functional enrichments additional functions should be specified. In relation to that topic it is valuable to point out that ordinary conditional equations in the context of partial algebras can be used as a complete functional language over arbitrary parameterized abstract data type. Thus the same calculus used to define the fundamental operations of parameterized abstract data types can be used to define arbitrary functional enrichments.

(4) Already the use of total operations and of equations leads to undecidable algebraic theories. Since an abstract data type should possess at least one model with recursively enumerable carrier sets and with partial recursive fundamental operations, one should avoid undecidable algebraic theories as specifications of abstract data types. However, there are no syntactical means to exclude undecidable algebraic theories. But point (3) provides a practicable tool. Roughly one can say that a specification defines a recursive abstract data type iff the equality can be specified by means of possibly partial auxiliary functions and of conditional equations. In addition, practical experiences show that the specification of the equality is very helpful for the detection of errors in a specification. The concept of a 'recursive abstract data type' will be extended to parameterized abstract data types in section 3 of this paper.

We hope that points (1) to (4) demonstrate reasons and advantages of the use of structural induction on partial algebras.

A second aim of the paper is the exemplification of the interaction of Universal Algebra and theoretical Computer Science. We are not able to present the whole range of that interaction here. A comprised but rather complete review of the interaction between theoretical computer science and universal algebra, model theory and category theory is given by Andreka and Nemeti in [AN '79] and [AN '80]. Recently very interesting results could be achieved by using concepts of abstract model theory (see [GB '83], [Mar '83] and [Tar '83].

1. Algebraic foundation of structural induction

The basic principle for defining a function $f:M \to A$ by structural induction is the definition of a signature Σ and a Σ-algebra with carrier A in such a way that M becomes the carrier of an initial Σ algebra and $f:M \to A$ becomes the unique homorphism from the initial algebra (see [ADJ '78]).
An extension of structural induction to partial algebras requires the possibility of defining sufficient and also necessary domain conditions. This can be done by so-called 'elementary implications' in the terminology of [BR '83] or by QE-equations as they are called by Burmeister (see [Brm '83], [ABN '80]). For a further algebraic approach to partial abstract data types see [BW '82].

Let $\Sigma = (S,(\Omega_{w,s} | w \in S^*, s \in S))$ be a signature, where S is the set of sort names and $\Omega = (\Omega_{w,s} | w \in S^*, s \in S)$ is the family of operation symbols (operators). For $\sigma \in \Omega_{w,s}$ we will also write $s:w \multimap s$ and we call $w \in S^*$ the domain of σ and $s \in S$ the range of σ.
A partial Σ-algebra

$$A = ((A_s | s \in S),(\sigma^A | \sigma \in \Omega))$$

is given by an S-indexed family of sets (the carrier of A) and by a many-sorted partial operation

$$\sigma^A : A_{s_1} \times A_{s_2} \times \ldots \times A_{s_n} \multimap A_{s_o}$$

for each operator $\sigma : s_1 s_2 \ldots s_n \multimap s_o$ in Ω.
A Σ-homomorphism $f:A \to B$ between partial Σ-algebras A,B is given by an S-indexed family of total functions.

$$f = (f_s : A_s \to B_s | s \in S)$$

such that for each $\sigma : s_1 s_2 \ldots s_n \multimap s_o$ in Ω and each $\underline{a} = (a_1, a_2, \ldots, a_n) \in \mathrm{dom}\sigma^A \subseteq A_{s_1 s_2 \ldots s_n}$ hold

(i) $\underline{a}\, f_{s_1 s_2 \ldots s_n} = (a_1 f_{s_1}, a_2 f_{s_2}, \ldots, a_n f_{s_n}) \in \mathrm{dom}\sigma^B$

(ii) $(\underline{a}\, f_{s_1 \ldots a_n})\sigma^B = (\underline{a}\sigma^A) f_{s_o}$

where the notation xf is used instead of $f(x)$.

Since we allow empty carriers we can not use a universal system of S-sorted variables. We use functions $u:X \to S$ as S-sorted systems of variables and define for any S-indexed family $A=(A_s | s \in S)$ the set A_u of all assignments of $u:X \to S$ in A by

$$A_u = \{\underline{a}: X \to \bigcup_{s \in S} A_s \mid x\underline{a} \in A_{xu}\} .$$

The problems resulting from possibly empty carriers can be solved by associating with every elementary implication an individual S-sorted system of variables (see [GM '81]).

For any S-sorted system of variables $u: X \to S$ the free term algebra of Σ-terms with variables of X will be denoted by

$$T(\Sigma, u).$$

Analogously to the operator symbols we will write $t: u \multimap s$ for $t \in T(\Sigma,u)_s$, $s \in S$. Each Σ-term $t: u \multimap s$ defines a partial term function

$$t^A: A_u \multimap A_s$$

for each partial Σ-algebra A.

An existence equation $(u: t_1 \overset{e}{=} t_2)$ with domain $u: X \to S$ is given by $t_1, t_2 \in T(\Sigma,u)_s$ for some $s \in S$. For any partial Σ-algebra A we denote with

$$A_{(u:\ t_1 \overset{e}{=} t_2)} = \{\underline{a} \in A_u \mid \underline{a} \in \text{dom } t_1^A \cap \text{dom } t_2^A,\ \underline{a}t_1^A = \underline{a}t_2^A\}$$

the set of all solutions of $(u: t_1 \overset{e}{=} t_2)$ in A.

For any set $(u:G) = (u: \{t_i \overset{e}{=} r_i \mid i \in I\})$ of existence equations with common domain $u: X \to S$

$$A_{(u:G)} = \bigcap_{i \in I} A_{(u:\ t_i \overset{e}{=} r_i)}$$

denotes the set of solutions of $(u:G)$ in A. For the empty set $(u:\emptyset)$ of existence equations we set $A_{(u:\emptyset)} = A_u$.

The basic tool of expressing properties of operations are elementary implications

(*) $(u: t_1 \overset{e}{=} r_1, \ldots, t_n \overset{e}{=} r_n \to t_o \overset{e}{=} r_o)$, $n \geq o$.

A partial Σ-algebra A satisfies (*) iff

$$A_{(u:\ t_1 \overset{e}{=} r_1, \ldots, t_n \overset{e}{=} r_n)} \subseteq A_{(u:\ t_o \overset{e}{=} r_o)} .$$

The following three special cases indicate the broad range of expressiveness of elementary implications.

$(u: \emptyset \to t_o \overset{e}{=} r_o)$, or $(u: t_o \overset{e}{=} r_o)$ for short, expresses that $\text{dom } t_o^A = \text{dom } r_o^A = A_u$ and that $\underline{a}\ t_o^A = \underline{a}\ r_o^A$ for each $\underline{a} \in A_u$ if A satisfies that elementary implication. Especially, t_o^A and r_o^A have to be total term functions.

The case

$$(u: t_1 \overset{e}{=} r_1, \ldots, t_n \overset{e}{=} r_n \to t \overset{e}{=} t)$$

represents a sufficient condition for the domain of the term t

and $(u: t \overset{e}{=} t \to t_1 \overset{e}{=} r_1), \ldots, (u: t \overset{e}{=} t \to t_n \overset{e}{=} r_n)$ represents a necessary condition for the domain of t.

In general an elementary implication expresses as well conditions of interrelations as of applicability of operations. For elementary implications a Compactness Theorem and a Completeness Theorem can be proved. The main result concerning structural induction is the existence of partial algebras freely generated by a set of generators and a set of defining relations.

Let Σ be any signature, $\mathfrak{A}$ any finite set of elementary implications and $(u:G)$ any set of existence equations (where $u:X \to S$ and G may be infinite). Then there exists up to isomorphism exactly one partial Σ-algebra $\underline{F}(\Sigma, \mathfrak{A}, (u:G))$ which satisfies each elementary implication of $\mathfrak{A}$ and which possesses a so-called universal solution

$$\underline{e} \in F(\Sigma, \mathfrak{A}, (u:G))_{(u:G)},$$

i.e., for any solution $\underline{b} \in B_{(u:G)}$ in any partial Σ-algebra B satisfying each elementary implication in $\mathfrak{A}$ there is exactly one homomorphism $f: F(\Sigma, \mathfrak{A}, (u:G)) \to B$ with $\underline{e} f_u = \underline{b}$.

Initial $(\Sigma, \mathfrak{A})$-algebras can now be constructed by $F(\Sigma, \mathfrak{A}, \emptyset)$, i.e. freely generated by the empty set of generators and by the empty set of defining relations.

2. Recursive abstract data types

An abstract data type is the isomorphism class of some $F(\Sigma, \mathfrak{A}, \emptyset)$. An abstract data type will be called recursive if the isomorphism class of $F(\Sigma, \mathfrak{A}, \emptyset)$ contains at least one Σ-algebra with recursively enumerable carrier sets and partial recursive fundamental operations. In that case we say that $(\Sigma, \mathfrak{A})$ initially specifies the abstract data type. Because of the compactness and the completeness theorem an initial algebra $F(\Sigma, \mathfrak{A}, \emptyset)$ can be built from recursively enumerables sets and partial recursive functions if and only if $(\Sigma, \mathfrak{A})$ is decidable. The following two theorems give the connections between free partial algebras and computability.

Theorem 1: $(\Sigma, \mathfrak{A})$ initially specifies a recursive abstract data type iff there is a finite extension $(\Sigma', \mathfrak{A}')$ of $(\Sigma, \mathfrak{A})$ containing a sort name $Bool \in sorts(\Sigma')$ and operators $\sigma_s: ss \multimap Bool$ for each $s \in sorts(\Sigma)$ such that

(1) the restriction of $F(\Sigma', \mathfrak{A}', \emptyset)$ to sorts and operators of Σ yields $F(\Sigma, \mathfrak{A}, \emptyset)$, i.e. $F(\Sigma', \mathfrak{A}', \emptyset)_s = F(\Sigma, \mathfrak{A}, \emptyset)_s$ for each $s \in sorts(\Sigma)$ and

$$\sigma^{F(\Sigma', \mathfrak{A}', \emptyset)} = \sigma^{F(\Sigma, \mathfrak{A}, \emptyset)}$$

for each $\sigma \in operators(\Sigma)$

(2) $(t_1, t_2)\sigma_s^{F(\Sigma', \mathfrak{A}', \emptyset)}$ = True iff $\mathfrak{A} \vdash t_1 \overset{e}{=} t_2$ for all $t_1, t_2 \in F(\Sigma, \quad, \emptyset)_s$, $s \in sorts(\Sigma)$, where $\mathfrak{A} \vdash t_1 \overset{e}{=} t_2$ denotes that $(u{:}\ t_1 \overset{e}{=} t_2)$ is syntactically derivable from

Additionally one can prove

Theorem 2: For any $(\Sigma, \mathfrak{A})$ initially specifying a recursive abstract data type and a partial function

$$\phi : F(\Sigma, \mathfrak{A}, \emptyset)_w \multimap\!\!\!\rightarrow F(\Sigma, \mathfrak{A}, \emptyset)_s$$

with $w:\{1,\ldots,n\} \to sorts(\Sigma)$, $s \in sorts(\Sigma)$, that function is partial recursive iff there is a finite extension $(\Sigma', \mathfrak{A}')$ of $(\Sigma, \mathfrak{A})$ satisfying condition (1) above and containing an operator $\sigma: w \multimap\!\!\!\rightarrow s$ with $\phi = \sigma^{F(\Sigma', \mathfrak{A}', \emptyset)}$.

The preceding theorems show that structural induction suffices for initial specifications of arbitrary recursive abstract data types and of arbitrary computable functions on recursive abstract data types. But, structural induction on partial algebras also allows initial specifications of data types whose carriers are only representable as quotient sets of recursively enumerable sets relative to a recursively enumerable equivalence relation. The theorems characterize specifications of recursive abstract data types as those that permit an initial specification of the identity.

3. Recursive dependence in initial canons

For an extension of Theorem 1 and Theorem 2 from abstract data types to parameterized abstract data types we need at first the extension of algebraic theories $(\Sigma, \mathfrak{A})$ to initial canons (or data theories) as defined in [Rei '80] and [BG '80].

An initial canon is given by an algebraic theory $(\Sigma, \mathfrak{A})$ and by a finite set Δ of initial restrictions of $(\Sigma, \mathfrak{A})$, where an initial restriction is roughly given by a chain of two subtheories $\delta = [(\Sigma_1, \mathfrak{A}_1) \subseteq (\Sigma_2, \mathfrak{A}_2) \subseteq (\Sigma, \mathfrak{A})]$ denoted $\delta = [T_1 \subseteq T_2 \subseteq T]$ for short.

The general notion of a canon allows that instead of the second inclusion a theory morphism is used.

We say that $A \in Alg(\Sigma, \mathcal{A})$ satisfies δ, written $A \models \delta$, if $A\downarrow(\Sigma_2, \mathcal{A}_2)$ is freely generated by $A\downarrow(\Sigma_1, \mathcal{A}_1)$, where $A\downarrow(\Sigma_i, \mathcal{A}_i)$ for $i=1,2$ denotes the restriction of A to sorts and operators of Σ_i that produces automatically an algebra $A\downarrow(\Sigma_i, \mathcal{A}_i) \in Alg(\Sigma_i, \mathcal{A}_i)$ for each $A \in Alg(\Sigma, \mathcal{A})$. Therefore, each $s \in sorts\,\Sigma_2 \setminus sorts\,\Sigma_1$ and each $\sigma \in operators\,\Sigma_2 \setminus operators\,\Sigma_1$ is subject to the initial restriction $\delta \in \Delta$. We say that $s \in sorts\,\Sigma$ or $\sigma \in operators\,\Sigma$ is restricted by Δ if it is subject to some $\delta \in \Delta$.

The range of model classes of canons expands from isomorphism classes to model classes of algebraic theories and covers the model classes of finite first order theories.

We illustrate the notion of canons by a specification of the semantics of a loop statement given by the following flowchart.

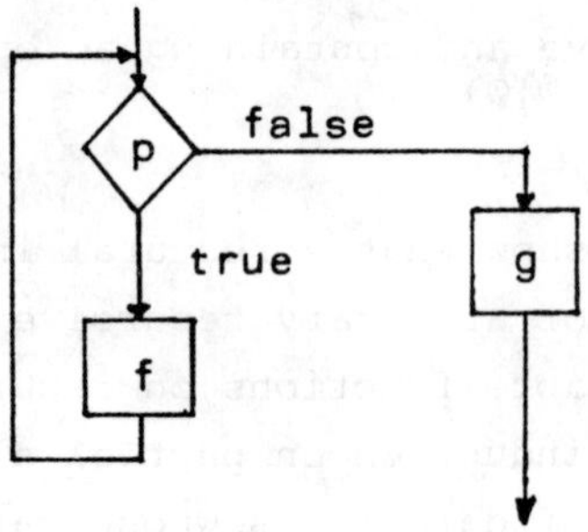

LOOP <u>is</u> <u>definition</u>

    <u>sorts</u> Bool
    <u>oprn</u> true, false ——> Bool   ------↓-- T_1
    <u>with</u> <u>requirement</u>
    <u>sorts</u> States
    <u>oprn</u> p(States) ——> Bool
        f(States) ——> States
        g(States) ——> States   ------↓-- T_2
    <u>with</u> <u>definition</u>
    <u>oprn</u> while(States) —○—> States
    <u>axioms</u> x:States
    <u>if</u> p(x)=false <u>then</u> while(x) = g(x)
    <u>if</u> p(x)=true, while(f(x))=while(f(x))
        <u>then</u> while(x) = while(f(x)) <u>fi</u> ------↓ T

<u>end</u> LOOP

Formally the preceding specification LOOP represents the canon
$LOOP = (T, \{\delta_1 = (\emptyset \leq T_1 \subseteq T),\ \delta_2 = (T_2 \leq T \subseteq T)\})$.

In a specification an arrow ——> indicates that the corresponding operators have to be interpreted by total operations and an arrow -∘—> indicates that the corresponding operator may be interpreted by a partial operation. According to that convention the subcanon $\mathbb{P} = (T_2, (\emptyset \leq T_1 \subseteq T_2))$ of the canon LOOP contains only total operators. The canonical axioms that cause the totality of the operators are omitted in a specification. The initial restriction $(\emptyset \leq T_1 \subseteq T_2)$ causes that for each $A \in Alg\mathbb{P}$ the two-element set $\{true, false\} = A_{Bool}$ serves as set of truth values. A_{States} may be any set because the sort States is not restricted in $\mathbb{P}$. $f^A, g^A : A_{States} \rightarrow A_{States}$ may be arbirtrary total functions and $p^A : A_{States} \rightarrow \{true, false\}$ may be any total predicate. Consequently $\mathbb{P}$ specifies the class of parameters of the loop-construction. The enrichment of $\mathbb{P}$ to LOOP by a definition causes that 'while' has to be interpreted precisely in the intended manner. This can easily be proved by induction. The (relative) initiality of $A \in AlgLOOP$ guarantees that $while^A$ is only defined if it can be deduced from the axioms. The first elementary implication requires

$$\{x \in A_{States} \mid p^A(x) = false\} \subseteq dom\ while^A$$

and the second axiom expresses the following domain conditon
'$p^A(x) = true$ and $f^A(x) \in dom\ while^A$ implies $x \in dom\ while^A$'
and additionally the following conditon of interrelation
'$p^A(x) = true$ and $f^A(x) \in dom\ while^A$ implies $while^A(x) = while^A(f^A(x))$'.
The following enrichment by definiton LOOP* of LOOP specifies a copy of the domain of 'while' as a new sort named 'Domain'.

```
LOOP* is LOOP with definition
      sorts Domain
      oprn d(States) —∘—> Domain
      axioms x:States
      if  while(x) = while(x) then  d(x) = d(x)
      end LOOP*
```

Intuitively the operator while(States) —∘—> States and the sort name 'Domain' represent recursive constructions over $\mathbb{P}$. For any interpretation of $\mathbb{P}$ by recursively enumerable sets and partial recursive functions we obtain a recursively enumerable interpretation

of 'Domain' and a partial recursive interpretation of 'while(States) —∘—> States'. This intuition leads to the following informal defintion:
Let $\mathbb{C}$ be any canon, $\mathbb{C}_o \subseteq \mathbb{C}$ any subcanon and $s_o \in sorts\, \mathbb{C}$. $s_o \in sorts\, \mathbb{C}$ is called <u>recursively dependent from $\mathbb{C}_o$</u> if the familiy $(IdA_{s_o} \mid A \in Alg\, \mathbb{C})$ is specifyable by an enrichment by definition of $\mathbb{C}_o$ extended by the s-identity for each $s \in sorts\, \mathbb{C}_o$. Each $s \in sorts\, \mathbb{C}$ recursively dependent from the empty subcanon is called <u>absolutely recursive</u>.
An operator $\sigma(s_1 \ldots s_n)$ —∘—> s_o in operators $\mathbb{C}$ is called <u>recursively dependent from $\mathbb{C}_o$</u> if each of $s_1, \ldots, s_n, s_o$ is recursively dependent from $\mathbb{C}_o$ and if the family

$$(\sigma^A \mid A \in Alg\, \mathbb{C})$$

is specifyable by an enrichment by definition of $\mathbb{C}_o$ again extended by the s-identity for each $s \in sorts\, \mathbb{C}_o$. Any operator of $\mathbb{C}$ is called <u>absolutely recursive</u> if it is recursively dependent from the empty subcanon of $\mathbb{C}$.

In the preceding example 'Bool', 'true', 'false' are absolutely recursive, and 'Domain' and 'while' are recursively dependent from $\mathbb{P} \subseteq LOOP^*$.

The concept of recursive dependence in canons is a useful notion for dealing with implementations of parameterized abstract data types by other abstract data types of lower conceptual level but with the same parameters. Such an implementation always maps a given data type (a sort name) onto a data type that is recursively dependent from the parameter subcanon. The application of recursive dependence to implementations of abstract data types in accordance with the CAT-concept of Burstall and Goguen [GB '79] is subject of a forthcoming paper. This application requires the extension of the concept of behavioural equivalence to data types with partial operations (see [Rei '81],[GM '82],[HR '83]). But in this case the partiality causes some serious algebraic problems which can not be explained here in detail.

The presented algebraic approach to recursion theory is a result of joint research work of the author with H. Kaphengst. The first version was presented in [KR '71]. Improved versions are described in [Rei '79] and [Kap '82].

References:

[ABN '80] Andreka, H., Burmeister, P. Nemeti, I.: Quasivarieties of partial algebras, Preprint Technische Hochschule Darmstadt, 1980; to appear in Studia Sci. Math. Hungar..

[ADJ '78] Goguen, J.A., Thatcher, J.W., Wagner, E.: An initial algebra approach to the specification, correctness and implementation of abstract data types. In R. Yeh (editor) Current Trends in Programming Methodology, Prentic-Hall, 1978.

[AN '79] Andreka, H., Nemeti, I.: Application of universal algebra, model theory, and categories in computer science. Computational Linguistics and Computer Languages, Vol. XIII.

[AN '80] Andreka, H., Nemeti, I.: Additions to survey of applications of universal algebra, model theory, and categories in compter science. Computational Linguistics and Computer Languages, Vol. XIV, 1980.

[Bur '69] Burstall, R.M.: Proving properties of programs by structural induction. Computer Journal, February 1969.

[BG '80] Burstall, R.M., Goguen, J.A.: The semantics of CLEAR, a specification language. Abstract Software Specifications, Lecture Notes in Computer Science, Vol. 86, Springer, 1980.

[BG '81] Burstall, R.M., Goguen, J.A.: Algebras, theories and freeness: an introduction for computer scientists. In Proc. of Marktoberdorf Summer School on Theoretical Foundations of Programming Methodology, August 1981.

[BR '83] Benecke, K., Reichel, H.: Equational partiality, Algebra Universalis, 16, pp. 219-232, 1983.

[Brm '82] Burmeister, P.: Partial algebras - survey of a unifying approach towards a tow-valued model theory for partial algebras. Algebra Universalis 15, pp. 306-358, 1982.

[BW '82] Broy, M., Wirsing, M.: Partial Abstract Types, Acta Informatica 18, pp. 47-64, 1982.

[GB '83] Goguen, J.A., Burstall, R.M.: Introducing Institutions Proc. Logics of Programming Workshop, CMU.

[GB '80] Goguen, J.A., Burstall, R.M.: CAT, a system for the structured elaboration of correct parograms from structured specifications. Technical Report CSL-118, Computer Science Laboratory, SRI International, 1980.

[GM '82] Goguen, J.A., Meseguer, J.: Universal realization, persistent interconnection and implementation of abstract modules. In ICALP' 82, Lecture Notes in Computer Science.

[GM '81] Goguen, J.A., Meseguer, J.: Completeness of many-sorted equational logic. SIGPLAN Notices 16, 7, pp. 24-32, 1981.

[HR '83] Hupbach, U.L., Reichel, H.: On behavioural equivalence of abstract data types. EIK 19, 6, pp. 297-305, 1983.

[KR '71] Kaphengst, H., Reichel, H.: Algebraische Algorithmentheorie. Wiss. Inf. und Berichte, Nr. 1, Reihe A, VEB Robotron, ZFT, Dresden 1971.

[Kap '81] Kahengst, H.: What is computable for abstract data types? In Proc. FCT' 81, Lecture Notes in Computer Science, Vol. 117, pp. 173-181.

[Mar '83] Makowsky, J.A.: Model Theoretic Issues in Theoretical Computer Science, Part 1: Relational Data Bases and Abstract Data Types, to appear in Logic Colloquium' 82, Ed. Lollo, G., Longo, G., Marcja, A., North Holland 1983.

[Rei '79] Reichel, H.: Theorie der Äquoide, Dissertaton B, Humboldt-Universität zu Berlin, 1979.

[Rei '80] Reichel, H.: Initially restricting algebraic theories. In Proc. MFCS '80, Lecture Notes in Computer Science, Vol. 88, pp. 504-514.

[Rei '81] Reichel, H.: Behavioural equivalence - a unifying concept for initial and final specification methods. In Proc. Third Hung. Compt. Sci. Conf., M. Arato, L. Varga (eds.), Akademiai Kiado, Budapest, 1981.

[Tar '83] Tarlecki, A.: Free constructions in algebraic institutions. Report CSR-149-83, Dept. of Computer Science, Univ. of Edinburgh.

H. Reichel
TH 'Otto von Guericke'
Sektion Matheamtik und Physik
DDR-3010 Magdeburg
Boleslaw Bierut Platz 5

List of participants

M. Albert, Oxford
P. Alles, Darmstadt
H. Andreka, Budapest
B. Artmann, Darmstadt
H. Bargenda, Stuhr
B. Bornscherer, Kassel
B. Banaschewski, Hamilton
M. Becker, Kaiserslautern
G. Backes, Kaiserslautern
H. Bauer, Darmstadt
A. Beutelspacher, Mainz*
B. Bosbach, Kassel
U. Brehm, Freiburg
G. Beuttenmüller, Ulm
W. Büttner, Darmstadt
P. Burmeister, Darmstadt
R. Castoral, Wien
P.M. Cohn, London*
G. Cupona, Skopje
E. Dahlhaus, Berlin
M. Dichtl, Ulm
K.H. Diener, Köln
H. Dobbertin, Hannover
J. Dörflinger, Ulm
G. Dorn, Darmstadt
J. Dudek, Wrocław
E. Ellers, Toronto
W. Engelmann, Darmstadt
U. Erbar, Kaiserslautern
L. Felsenstein, Darmstadt
C.C. Ferrero, Parma
W. Fey, Berlin
U. Faigle, Bonn
W. Felscher, Tübingen*
A. Flörke, Darmstadt
E. Fried, Budapest*
W. Frisch, Kaiserslautern
B. Ganter, Darmstadt
O. Garcia, Mexico
H. Gensheimer, Ulm
H. Gerstmann, Hannover
R. Godowski, Warszawa
G. Grätzer, Winnipeg*
H.J. Groh, Darmstadt
G. Grimeisen, Stuttgart
A. Gülzow, Kassel
I. Guessarian, Paris
H.P. Gumm, München*
G. Hansoul, Liege
R. Harting, Düsseldorf
E. Harzheim, Düsseldorf
J. Hagemann, München
H. Hansen, Berlin
F. Hergert, Darmstadt
C. Herrmann, Darmstadt
A. Herzer, Mainz*
W. Hodges, London*
K.H. Hofmann, Darmstadt
R.E. Hoffmann, Bremen
E. Hotzel, Bonn
T. Ihringer, Darmstadt
R. John, Bonn
H. Jürgensen, Darmstadt*
H.K. Kaiser, Wien*
G. Kalmbach, Ulm
H. Kautschitsch, Klagenfurt
K. Keimel, Darmstadt
P. Köhler, Giessen
H. König, Erlangen
O. Kopecek, Stuttgart
G. Kowol, Wien

* Invited lecturers

P. Krauss, Kassel*
E. Kronz, Kaiserslautern
U. Kipke, Darmstadt
O.H. Kegel, Freiburg
H. Kröger, Kiel
H. Länger, Wien
H. Lenzing, Paderborn
W. Lex, Clausthal-Zellerfeld
I. Lienkamp, Darmstadt
P. Luksch, Darmstadt
S. Mac Lane, Chicago*
S. Markowski, Skopje
H. Mäurer, Darmstadt
L. Marki, Budapest
R. Mlitz, Wien
R. Metz, St. Augustin
K.D. Meyer, Kaiserslautern
H. Mitsch, Wien
A. Mitschke, München
R.H. Möhring, Aachen*
H. Möller, Eppstein
I. Molnar, Budapest
E. Müller, Kassel
W. Müller, Klagenfurt
W. Müller, Darmstadt
P. Nevermann, Darmstadt
I. Nemeti, Budapest*
R. Nausester, Darmstadt
W. Nolte, Darmstadt
G. Oehme, Darmstadt
P. Padawitz, Berlin
A. Pasztor, Stuttgart
M. Paul, Kassel
S. Pellegrini, Parma
E. Pesch, Darmstadt
G. Pilz, Linz
W. Poguntke, Darmstadt
A. Poigne, Dortmund
F. Poyatos, Madrid
H. Priestley, Oxford
R. Quackenbush, Winnipeg*
H. Ratscheck, Düsseldorf
H. Reichel, Magdeburg*
K. Reuter, Darmstadt
M. Richter, Aachen*
G. Richter, Bielefeld
A. Romanowska, Warszawa
W. Ruckelshausen, Darmstadt
J. Schäfer, Darmstadt
A. Schlegel, Darmstadt
J. Schmid, Bern
E.T. Schmidt, Budapest*
P.H. Schmitt, Heidelberg
J. Schulte-Mönting, Tübingen
D. Schweigert, Kaiserslautern
D. Scott, Pittsburgh*
K. Siefert, Ulm
M. Siegmund-Schultze, Darmstadt
U. Simon, Saarbrücken
J. Smith, Darmstadt
G. Snelting, Darmstadt
H. Sperber, Erlangen
E. Stachow, Köln
J. Stahl, Darmstadt
M. Steinby, Turku
M. Stone, Calgary
K. Strambach, Erlangen*
H. Strehl, Erlangen
A. Suppa, Parma
D. Suter, Mannheim

W. Sydow, Darmstadt

L. Teschke, Saarbrücken

V. Trnkova, Prag*

B. Voigt, Bielefeld

A. Ursini, Siena

J. Varlet, Liege

H. Volger, Tübingen

L. Vrancken-Mawet, Belgien

V. Weispfenning, Heidelberg*

G. Wenzel, Mannheim

H. Werner, Kassel*

R. Wille, Darmstadt

K.E. Wolff, Darmstadt

E. Wronska, Opole

P. Zahn, Darmstadt